# Shellfish & Seaweed Harvests Of Puget Sound

# Shellfish & Seaweed Harvests Of Puget Sound

Daniel P. Cheney · Thomas F. Mumford, Jr.

**Washington Sea Grant Program**
Distributed by the University of Washington Press
Seattle and London

First Published in 1986 by
Washington Sea Grant Program
University of Washington

Distributed by University of Washington Press
Seattle, Washington 98195



Publication of this book was supported by grants (04-5-158-48; 04-7-158; NA79AA-D-00054, NA81AA-D-00030, and NA84AA-D-00011) from the National Oceanic and Atmospheric Administration and by funds from the Environmental Protection Agency. Writing and publication was conducted by the Washington Sea Grant Program under project A/PC-7.

**Library of Congress Cataloging-in-Publication Data**

Cheney, Daniel P., 1941-
Shellfish and seaweed harvests of Puget Sound.

(Puget Sound books) (A Washington sea grant publication)
Bibliography: p.
Includes index.
1. Shellfish fisheries—Washington (State)—Puget Sound. 2. Marine algae culture—Washington (State)—Puget Sound. 3. Marine algae—Washington (State)—Puget Sound. I. Mumford, Thomas F., 1944-
II. Washington Sea Grant Program. III. Title.
IV. Series. V. Series: Washington sea grant publication.
SH365. W2C48 1986 338.3'72594'11097977 85-17913
ISBN 0-295-95990-8

# Contents

To the memory of Ivar Haglund

Whose love of Puget Sound was an inspiration to

all those who live in the Pacific Northwest

**Project Staff**

Alyn Duxbury
Director

Patricia Peyton
Managing Editor

Andrea Jarvela
Editor

Kirk Johnson
Designer

**Production Staff For this Volume**

Andrea Jarvela
Margaret Darland
Editors

Constance Bollen
Designer/Illustrator

Vicki Miles
Illustrator

**Supporting Staff**

Laura Mason



Funds to support the publication of the Puget Sound Books were provided by the National Oceanic and Atmospheric Administration (NOAA) and by the Environmental Protection Agency (EPA).

# About the Puget Sound Books

This book is one of a series of books that have been commissioned to provide readers with useful information about Puget Sound . . .

**About its physical properties**—the shape and form of the Sound, the physical and chemical nature of its waters, and the interaction of these waters with the surrounding shorelines.

**About the biological aspects of the Sound**—the plankton that form the basis of its food chains; the fishes that swim in this inland sea; the region's marine birds and mammals; and the habitats that nourish and protect its wildlife.

**About man's uses of the Sound**—his harvest of finfish, shellfish, and even seaweed; the transport of people and goods on these crowded waters; and the pursuit of recreation and esthetic fulfillment in this marine setting.

**About man and his relationships to this region**—the characteristics of the populations which surround Puget Sound; the governance of man's activities and the management of the region's natural resources; and finally, the historical uses of this magnificent resource—Puget Sound.

To produce these books has required more than six years and the dedicated efforts of more than one hundred people. This series was initiated in 1977 through a survey of several hundred potential readers with diverse and wide-ranging interests.

The collective preferences of these individuals became the standards against which the project staff and the editorial board determined the scope of each volume and decided upon the style and kind of presentation appropriate for the series.

In the Spring of 1978, a prospectus outlining these criteria and inviting expressions of interest in writing any one of the volumes was distributed to individuals, institutions, and organizations throughout Western Washington. The responses were gratifying. For each volume no fewer than two and as many as eight outlines were submitted for consideration by the staff and the editorial board. The authors who were subsequently chosen were selected not only for their expertise in a particular field but also for their ability to convey information in the manner requested.

Nevertheless, each book has a distinct flavor—the result of each author's style and demands of the subject being written about. Although each volume is part of a series, there has been little desire on the part of the staff to eliminate the individuality of each volume. Indeed, creative yet responsible expression has been encouraged.

This series would not have been undertaken without the substantial support of the Puget Sound Marine EcoSystems Analysis(MESA) Project within the Office of Oceanography and Marine Services/Ocean Assessment Division of the National Oceanic and Atmospheric Administration. From the start, the representatives of this office have supported the conceptual design of this series, the writing, and the production. Financial support for the project was also received from the Environmental Protection Agency and from the Washington Sea Grant Program. All these agencies have supported the series as part of their continuing efforts to provide information that is useful in assessing existing and potential environmental problems, stresses, and uses of Puget Sound.

Any major undertaking such as this series requires the efforts of a great many people. Only the names of those most closely associated with the Puget Sound Books—the writers, the editors, the illustrators and cartographers, the editorial board, the project's administrators and its sponsors—have been listed here. All these people—and many more—have contributed to this series, which is dedicated to the people who live, work, and play on and beside Puget Sound.

Alyn Duxbury and Patricia Peyton
August 1985

# Preface

This book was first conceived in 1977 and the first draft written in 1978. It was originally to be part of a large volume that included finfish, but that is to be a separate book. This book has been rewritten several times in order to reflect the many changes that have occurred in the shellfish and seaweed industries. We have seen the rapid expansion of new aquaculture crops, the development of shellfish hatcheries, and the beginnings of nori aquaculture in Puget Sound. We have also seen a heightened awareness and understanding of the fragile nature of the Sound's water quality.

The intent of this book is to describe shellfish and seaweed capture fisheries and aquaculture—past and present. It focuses on the commercial sector and the research, development, and political activities that have influenced these industries.

The information presented comes from a variety of sources—published and unpublished literature, many interviews, and our own experiences. While much of the information points toward trends, what we see and say today may not hold true for any length of time. We have no crystal ball. Products and techniques for raising or catching marine species of which we have no inkling today will someday undoubtedly become commonplace.

We hope this book will be of interest to those consuming seafood products from Puget Sound. We also hope it will be of value to those making political decisions that will influence our ability to harvest and culture such products in the future and that those contributing to such decisions will be the harvesters, growers, and consumers.

Strait of Georgia
Bellingham
Bellingham Bay
Vancouver Island
Haro Strait
Wescott Bay
Friday Harbor
San Juan I.
Cypress I.
Samish Bay
Skagit River
Anacortes
Victoria
American San Juan Islands
Whidbey I.
Skagit Bay
Strait of Juan de Fuca
Stillaguamish River
Penn Cove
Race Lagoon
Camano I.
Dungeness Spit
Dungeness Bay
Port Susan
Port Angeles
Sequim
Port Townsend
Sequim Bay
Miller Peninsula
Mission Beach
Everett
Useless Bay
Snohomish River
Quilcene Bay
Quilcene River
Port Gamble
Puget Sound
Appletree Cove
Olympic Peninsula
Dabob Bay
Liberty Bay
Agate Pass
Dyes Inlet
Winslow
Bainbridge I.
Elliott Bay
Seattle
Bremerton
Sinclair Inlet
Hood Canal
Duwamish River
Vashon I.
Burley Lagoon
Minter Bay
Oakland Bay
Case Inlet
Carr Inlet
The Narrows
Tacoma
Shelton
Hammersley Inlet
Hartstene I.
McNeil I.
Tillow Beach
Puyallup River
Skookum Inlet
Squaxin I.
Drayton Passage
Anderson I.
Totten Inlet
Oyster Bay
Eld Inlet
Budd Inlet
Henderson Inlet
Olympia
Nisqually River

# Acknowledgments

We would like to thank all those in the seafood industry who have given us much of the information needed for this book. Especially, we would like to thank Dick Steel, Gregg Bonacker, Earl Brenner, Jerry Yamashita, Peter Jefferds, and Bill Webb.

The Washington State Department of Fisheries and Natural Resources kindly provided access to much information and reviewed portions of the text. We would especially like to thank Dick Bumgarner, Lynn Goodwin, Bruce Pease, Ron Westley, and Cedric Lindsey. Jack Lilya of the Shellfish Sanitation Division of the Department of Social and Health Services and Bob Saunders of Department of Ecology were helpful in providing up-to-date information on the pressing issues of water quality. John Glude and Al Sparks provided much inspiration for the shellfish portion and provided valuable insight into the historical literature. There are many others, and because they are not named does not lessen their contributions. We have learned a great deal about these resources from everyone, and for that we are most grateful.

Without the perserverence and understanding of Andrea Jarvela Corell with these two first-time authors, and the efforts of Constance Bollen and Alyn Duxbury, this book would never have been completed.

CHAPTER ONE

# Out of an Inland Sea

Long before European explorers sailed into the inland waters of Puget Sound, coastal Indian tribes gathered and used its abundant native shellfish and seaweed resources. They collected oysters and clams for food and trade, cooled sunburned lips with sea lettuce, cured the stipes of bull kelp to use for fishing lines, and used the hollow kelp bulbs as storage containers. Even inland tribes periodically journeyed to the Sound to share this bounty. The evidence of this rich harvest remains in the shell piles—or "kitchen middens"—of coastal native village sites found along the shores of Washington State.

Yet there was no large-scale midwater harvest of crab or shrimp and no cultivation of oysters and clams by these Native Americans. Indians collected wild native oysters from natural beds, and later white settlers caught crabs from the intertidal areas, and fished for shrimp from the shore using beach seines. Tidelands were public property and anyone could dig for clams or collect oysters for personal use or to sell.

In 1890, the first Washington State Legislature passed legislation that would shape the future of Washington's intertidal shellfish industries. Known as the "Callow Act," this legislation provided for the sale of tidelands supporting natural oyster beds to private citizens. To further encourage the development of an oyster industry by assuring access to a husbanded crop, the state charged low prices for these lands but retained a reversionary right to reclaim lands that were used for purposes other than oyster cultivation.

In 1895, the legislature passed a second law, the "Bush Act," governing oyster lands. The Bush Act gave any citizen the right to file on oyster land, whether or not oysters had been previously cultivated there. However, it too specified that if the land was used for purposes other than oyster cultivation, the deed could be cancelled and the land revert to the state for resale. Furthermore, if the land was found to be unfit for oyster cultivation, the purchaser could cancel the deed and claim a different parcel of land.

---

*Gathering abalones*, photo by Edward S. Curtis, c. 1914. (Courtesy Special Collections Division, University of Washington Libraries)

As a result of the Bush and Callow Acts, much of the tidelands within Washington's bays and estuaries came under private ownership or lease, affecting not only harvest of wild oysters by the general public, but clams as well. The area of public tidelands in Puget Sound declined (approximately 40 percent is left), and as a result, the state has ceased tideland sales and stopped issuing commercial licenses to dig bay clams on publicly owned state tidelands, but continues to license clam-farming operations.

The development of new culture and harvest technologies, improved management, and the expanded use of other shellfish and plant species have opened up additional areas and resources in Puget Sound. For example, in the subtidal substrate, geoduck clams are now hand harvested by divers and subtidal hardshell clams are obtained using mechanical-hydraulic harvesting machines. Soft-shelled clams are sufficiently abundant in the intertidal zones of some embayments to warrant harvest. In addition, new advancements in artificially seeding shellfish beds have allowed an increase in yields beyond what is possible with recruitment from natural setting. Both the capture fishery, or wildstock harvest of shellfish and plants, and aquaculture or aquatic farming, have benefitted from these changes. The result is that wide ranges of shellfish and plant products grown in Puget Sound are available.

Puget Sound waters offer extensive possibilities for future aquaculture developments or for expansion of existing fisheries. The factors that have molded Puget Sound's traditional fishing industry—availability of suitable habitat, water quality, biology, variability of native stocks, changing markets, conflicting uses, and management practices—will continue to influence the development of new and existing fisheries.

## Physical Setting

Puget Sound's marine environment is varied and dynamic, with seasonal and geographic extremes in physical, chemical, and biological characteristics of water and a varied bottom topography. The shape and setting of the Sound influence the amount of freshwater present, currents, tidal fluctuations, wave action, and bottom properties, which in turn affect the distribution of organisms that exist in the various habitats of these waters. Distribution patterns exist because some shellfish and plants require lower salinity, some require coarser substrate, and some require a higher temperature to induce spawning.

The shape of Puget Sound was influenced by its geographical setting, located between the Olympic and Cascade mountain ranges, and the scouring that occurred from repeated glacial expansion and retreat. This scouring formed large, steep-sided, deep, narrow inlets with major

Olympia oyster beds were prevalent throughout south Puget Sound in the early 1900s. Shallow waters of quiet inner bays were intensively cultivated and diked to enhance growth and natural setting. Only scattered remnants of the diked beds remain. (Photo courtesy Special Collections Division, University of Washington Libraries, Negative 308)

basins separated by shallow sills. As sediments from river runoff and coastal erosion filled nearshore areas, large estuarine flats and subtidal basins formed near the mouths of major rivers and in bays and inlets. Shallow embayments influenced by the surrounding land have larger ranges of temperature and salinities than the open sea. Wetlands and shallow bays scattered throughout the Sound offer protected habitat for large populations of finfish, shellfish, and algae.

## Water Quality

Puget Sound is, for the most part, a well-mixed estuary. The presence of topographically rough sills and rapid tidal currents creates turbulence that mixes nutrient-rich deep water with the nutrient-poor, less saline surface waters. It is only during periods of high runoff that unmixed surface lenses of freshwater are found near the rivers that empty into the Sound.

Mixing of salt- and freshwater is further facilitated by net exchange of inland waters with oceanic seawater, which is typical of an estuarine system where diluted seawater moves seaward at the surface and ocean water moves inward at depth. This net exchange helps produce an extremely productive seaweed and phytoplankton growth that supports an abundance of marine life. In addition, the shallow embayments and wetlands of the estuaries, enriched by nutrients from streamflow and groundwater and decaying plant matter, offer habitat for commercially important finfish and shellfish.

Numerous biological and man-induced water quality factors affect the reproduction, growth, survival and utility of plants and shellfish in

Puget Sound. These range from the natural seasonal variations in plankton and detritus that comprise the food for many shellfish, to urban wastes. Industrial and domestic wastes can be highly toxic, particularly to sensitive species such as the Olympia oyster, or render exposed seafoods unfit for human consumption.

Certain natural features of Puget Sound waters preclude development of some fisheries, however. Species requiring high salinity water, such as the abalone, are generally restricted to the more oceanic waters of the San Juan Islands and Strait of Juan de Fuca, far from freshwater sources. In a few parts of Puget Sound—the mainstream of Hood Canal, for example—surface water temperature, nutrients, and salinity can be very low as spring snowmelt joins the surface layer. The presence of this freshwater later in the summer can cause high surface temperatures as the stratification helps limit vertical mixing.

Although water quality in Puget Sound is better than in many other major estuaries in the United States, pollutants have a major impact on fisheries and aquaculture. Shellfish harvests from the Sound's most productive waters have been banned, restricted, or threatened because of high levels of sewage bacteria coming from sewage treatment plants, improperly placed or failed septic tank systems, recreational boats and other vessels, hobby farms, and stormwater runoff. Toxic wastes, some of which are extremely poisonous to marine plants and animals and are common in marine sediments near industrial areas, have been discovered with disturbing regularity—but not always at toxic levels—in sediments at less developed locations in the Sound. In the chapters that follow these pollutant problems and others are discussed in greater detail as they affect specific marine harvests.

A significant water quality factor affecting shellfish in Puget Sound is paralytic shellfish poisoning (PSP). For many years, it has been known that humans can suffer poisoning by eating molluscan shellfish. It was not until the late 1920s that what is now termed "paralytic shellfish poisoning" was determined to be caused by large, seasonal concentrations of a toxic, free-swimming, marine dinoflagellate called *Gonyaulax*, which, when ingested by filter-feeding molluscs, resulted in accumulation of toxins in the body of the shellfish.

PSP toxin is potent: it can cause tingling and burning sensations in the face and extremities, numbness, muscular paralysis, and, in severe cases, death. Cooking does not remove the danger of PSP, but the percentage of toxin in the tissues can be greatly reduced by cooking shellfish at high temperatures. The only reliable way to determine whether the toxin is present at levels sufficient to be a health hazard is through laboratory bioassay that tests the toxicity on mice, although chemical procedures are being developed to replace bioassays as a detection method.

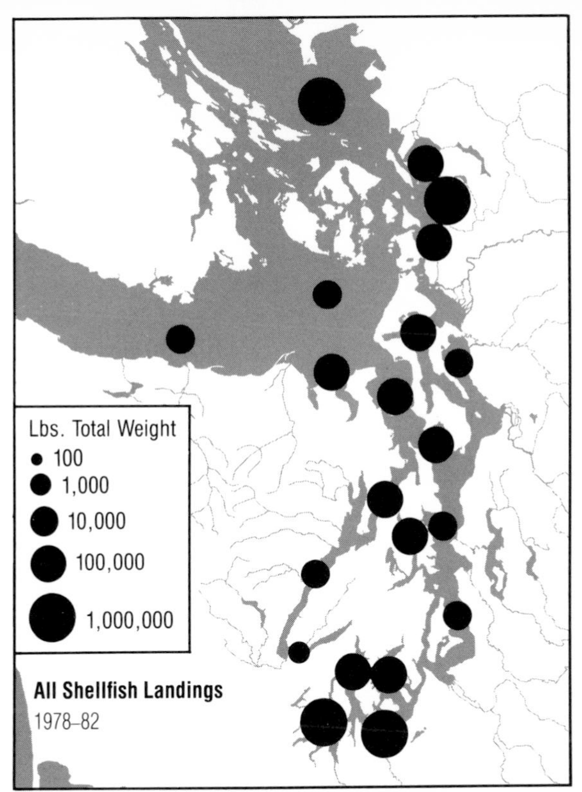


Distribution of shellfish landings in Puget Sound, 1978–82. Each circle depicts average annual landings from the value given up to the next highest value.

Until 1978, PSP contamination was confined largely to sporadic outbreaks in the Strait of Juan de Fuca, and although *Gonyaulax* was present in Puget Sound, it did not occur in numbers sufficient to produce toxicity in shellfish. However, in 1978 there was a major outbreak of PSP affecting shellfish as far south as central Puget Sound. Blue mussels and butter clams were the most severely affected, and most intertidal commercial and sport clam digging was banned. Since 1978, annual outbreaks of varying severity ("blooms") of *Gonyaulax* have resulted in restrictions on the harvest of clams, mussels, and other shellfish at numerous locations in north and central Puget Sound. Although these restrictions have not had a major impact on the commercial harvests, there is concern that PSP contamination could spread to south Sound. The cause of the outbreaks is unknown, and there is no evidence as yet to link them to any man-induced factor.

## Biology

Puget Sound supports a rich and varied assemblage of marine organisms, most of which are resident species that migrate only infrequently to coastal or offshore oceanic waters. In addition, almost all are either entirely marine or complete their life cycles in the marine environment. These biological systems as a whole are relatively stable; however, individual species may exhibit marked changes in abundance, and most have seasonal patterns of growth and reproduction.

Population levels and distribution of some species are limited by habitat availability and not by limitations in reproduction. For example, molluscs and crustaceans, are particularly dependent on the availability of appropriate habitat space and food of bays and estuaries. Also intimately linked with the physical and chemical factors of the environment are biological elements of reproduction and growth: spawning season, growth rate, time of spawning, developmental period, feeding habits, competition, predation, and disease.

## Conflicting Uses

Puget Sound's commercial shellfish and developing sea vegetable industries require natural resources that are also used for other purposes—recreation, onshore development, and municipal and industrial waste disposal are examples. In the Duwamish estuary, for example, historically valuable shellfish fisheries were destroyed when the tideflats were filled, the river channel dredged, and the area developed for industrial use. In many cases encroachment on marshes and tidelands has resulted in loss of biological productivity or loss of access to the site by fishermen.

Although shellfish and sea vegetable production may be increased by altering habitats, exploiting new harvest grounds, or employing intensive cultivation techniques, these procedures are sometimes in conflict with established environmental requirements in coastal waters, other uses, and aesthetic values of the area's residents. Conflicts in use have been most serious for molluscan species like oysters and clams, which occupy or are dependent upon shallow tideflats or subtidal regions. Organism preference for these sites and harvesting costs result in harvesting being conducted close to shore and often in proximity to upland dwellers. This activity has, in some instances, been considered by upland owners as degrading to the coastal environment. As a consequence, harvest or culture outside of traditional fishery areas is closely monitored and may be severely restricted.

## Markets

Because of high biological productivity, diversity of aquatic resources, and protection from oceanic storms, Puget Sound has played a major role in the commercial and recreational shellfishing industry on the West Coast. Shellfish are sold in restaurants and fish markets in Western Washington and in regional and national markets. Besides processing locally caught species, many local firms also process and market shellfish caught in waters off the Washington coast and in

Alaska. The total pounds landed and dollar value of shellfish harvests in Puget Sound are lower than those for finfish; however, the per unit value or price per pound is almost without exception much greater.

Crabs, oysters, clams, and mussels are the mainstays of the Puget Sound shellfish industry, but there are other abundant shellfish species, such as shrimp, squid, octopus, and sea urchins, that have a seasonal local market, are marketed outside of the United States, or are used primarily for bait. The premium prices paid in Japan for Washington State sea urchin gonads and use of the fast-growing octopus as bait by halibut fishermen demonstrate that potential markets exist for these underused native species.

In the past, consumer demand in the Puget Sound region has been fickle with regard to marine products such as sea vegetables, mussels, and squid. However, domestic and foreign markets for these items have expanded rapidly and some of them command high prices. Thus, there is potential for major expansion and sustained growth of commercial shellfisheries and shellfish and sea vegetable aquaculture in Puget Sound.

## Management

The commercial fisheries of Puget Sound supply a moderate but consistent yield of shellfish and a potential yield of sea vegetables. Traditionally, fishing pressure and yields followed abundance of species, whose population levels vary from year to year because of different growth and reproduction rates governed by natural variability in environmental conditions. To reduce the possibility of overexploitation

Clams, oysters, mussels, and crab comprise the core of Puget Sound's shellfish industry. These and small quantities of other types of shellfish are marketed fresh and sometimes live in local markets and throughout the United States.

Table 1.1 The missions of key government agencies in harvest and aquaculture management and regulation of shellfish and sea vegetables. (After Bish, 1982)

E Enhancement R Regulation P Protection

| Federal Agencies | Aquaculture | | | Capture Fishery | | | Authority or Role |
|---|---|---|---|---|---|---|---|
| Coast Guard | | R | P | | R | P | Responsible for marine safety, pollution abatement, channel marking and enforcement of maritime laws. |
| Corps of Engineers | | R | P | | | | Regulates construction in navigable waters (Section 10) and grants dredge and fill (Section 404) permits. |
| Environmental Protection Agency | | R | P | | | | Oversees Section 404 permits and toxics (Superfund) studies; point source discharges from aquaculture projects. |
| Food and Drug Administration | | R | | | R | | Enforces Food, Drug and Cosmetic Act; sets federal guidelines for food safety, shellfish beds, and processors. |
| Fish and Wildlife Service | E | R | P | E | | P | Has key role in NEPA and Fish and Wildlife Coordination Act (with EPA, COE and NMFS). |
| National Maritime Fisheries Service | E | R | P | E | R | P | Participates in management of the capture fishery; has an extensive research program and a role in NEPA. |
| Department of Agriculture | E | | | | | | Promotes aquaculture through agriculture programs; offers support important in protection of water quality. |
| Office of Ocean and Coastal Resource Management | | | P | | | P | Administers the Costal Zone Management Act—protects coastal resources via state CZM programs. |
| **Washington State Agencies** | | | | | | | |
| Department of Ecology | | R | P | | R | P | Administers CZM program through Shoreline Management Act; grants point source permits; conducts water quality tests. |
| Department of Fisheries | E | R | P | E | R | P | Issues hydraulics (HPA) permits (with DOG); regulates capture fishery; conducts research and enhancement. |
| Department of Game | | R | P | | R | P | Shares HPA authority with DOF; permits sale of game fish; oversees permitting in cases which may affect wildlife. |
| Department of Natural Resources | E | R | P | E | R | P | Conducts subtidal clam (with DOF) and seaweed research and development; administers leases of public aquatic lands. |
| Parks and Recreation Commission | | | P | | | P | Controls access to public beaches; administers (with DNR) tidelands abutting parks lands. |
| Dept. of Social and Health Services | | R | P | | R | P | Inspects and certifies shellfish beds for commercial harvest; conducts bacteriological and PSP monitoring. |
| Department of Agriculture | E | R | P | | | | Promotes aquaculture through agriculture programs, and cooperative support programs with the industry. |
| **Local Governments** | | | | | | | |
| Health Departments | | | P | | | P | Monitor water quality (with DOE), inspect (in some cases) on-site treatment systems. |
| County Commissioners/Planning Depts. | | R | P | | R | P | Formulate and implement land use plans, especially local zoning and shoreline management plans. Authority under SMA. |
| Public Works Departments | | R | P | | | P | Permit and inspect new construction usually including on-site treatment systems. |
| Port Districts | E | | | E | | | Authority parallels DNR for management of lands abutting uplands owned or controlled by district. |
| Shellfish Protection Districts | | | P | | | | Districts can establish special use areas and measures to protect resources from water quality degradation. |
| **Puget Sound Tribal Jurisdictions** | | | | | | | |
| Within Reservations | E | R | P | E | R | P | Primary authority for all activities rests with tribes unless permitted use is not consistent with CZM. |
| Outside Reservations | | | | E | R | P | Indirect regulation through NEPA and other federal programs. |

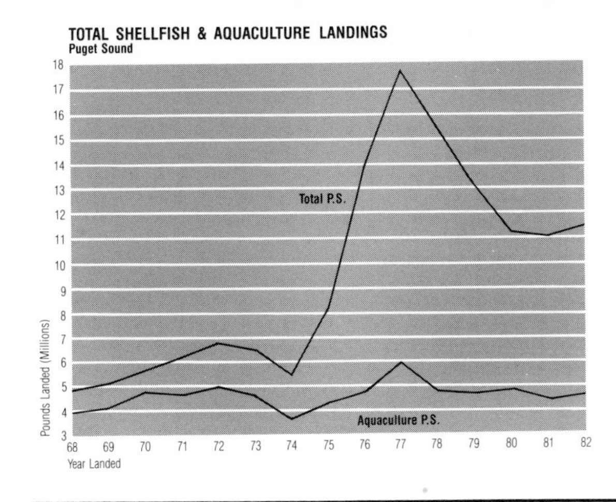


Total wild-capture harvest and aquaculture-produced landings of commercial shellfish from Puget Sound. The expanded fishery for geoduck has accounted for most of the increased production since 1971. (Source: Annual landings data for 1969–82, Washington Department of Fisheries)

when recruitment is low, government agencies have initiated regulatory procedures designed to control fishing pressure. In addition, some stocks—notably oysters and clams—are supplemented by controlled cultivation, sometimes by seeding with artificially reared juveniles to maintain harvest levels in the face of unpredictable natural reproduction.

To increase production of the varied shellfish stocks, management efforts are usually directed toward inducing changes in abundance, survival, or reproduction; substrate or sediment grain size composition; and appearance, quality, or catchability of the target species. This is a complex process because many species have overlapping environmental requirements. Actions that may enhance a particular family of shellfish or sea vegetable, a species area, or one segment of the fishing public may be detrimental to species composition, abundance, and distribution of other plants or animals in the marine community.

Commercial fishing is managed through a variety of regulations that directly or indirectly control when, where, how, and with whom a fisherman can fish. The Washington State Department of Fisheries has primary resource management responsibility for fisheries. In addition, there are many other agencies whose actions may profoundly affect part or all of the people involved in commercial and recreational fishing (Table 1.1). In order to keep discussions of species, their needs, potential, and regulation within reasonable limits, the chapters that follow concentrate on the commercial harvest of shellfish and seaweeds in Puget Sound and species that are or have been harvested here, or that have market potential for harvest or cultivation using proven aquaculture operations. Included are a number of edible invertebrate species that do not appear to have a market in the United States in the foreseeable future, but are native to or could be cultured in Puget Sound waters in sufficient numbers to export.

An oyster shucking line in the early 1920s. Except for the absence of conveyor belts and stainless steel working areas, this operation differs little from some modern facilities. (Photo courtesy Special Collections Division, University of Washington Libraries)

This float photographed in the late 1940s in Mud Bay holds egg crates used as cultch (substrate) to collect Olympia oyster spat or very young oysters. (Photo © Bob and Ira Spring)

CHAPTER TWO

# Oysters

The story of oystering in Puget Sound begins with the Olympia oyster, a small delicious creature that is the only edible oyster native to this region. It was the Olympia that accounted for the large naturally occurring oyster populations that once flourished in Puget Sound and Willapa Bay. The Olympia oyster's present abundance is only a fraction of historic levels, because it was extensively harvested in the late 1800s to supply a lucrative market in California. Commercial harvest of oysters in Washington State began as a result of the California gold rush in 1849. The gold brought new prosperity and money to pay for luxuries such as oysters shipped from Willapa Bay (then known as Shoalwater Bay).

Beginning with the first shipment in 1851 that arrived in San Francisco via schooner, the harvest and export of oysters expanded rapidly, from about 30,000 bushels in 1860 to a peak of over 130,000 bushels in the 1890s. Yet a thriving industry had no sooner become established when it faced stiff competition from large Eastern oysters, which were introduced to California after completion of the transcontinental Central Pacific Railroad in 1869.

East Coast oysters soon accounted for nearly 85 percent of the market in California, and during the turn of the century, demand for the little Olympias fell dramatically until contamination in East Coast shellfish caused a health concern among consumers. The Olympia oyster market rebounded, but the nature of the oyster business in Washington was already changing. Oyster beds in Willapa Bay had not been restocked or rehabilitated to their former levels, and in Puget Sound the gathering of wild stocks slowly began to evolve into the harvest of cultivated stocks. Puget Sound oystermen adapted a method of farming oysters using an elaborate system of diked ponds, a system that originated in France. With the water level stabilized behind the dikes, the oysters were no longer left exposed during low tide to freezing temperatures in winter and overheating in summer.

Diking of tidelands proved to be an effective way to enhance the growing environment for Olympia oysters in Puget Sound, but its adoption here was the result of important economic factors as well as environmental considerations. The building and maintenance of dikes was

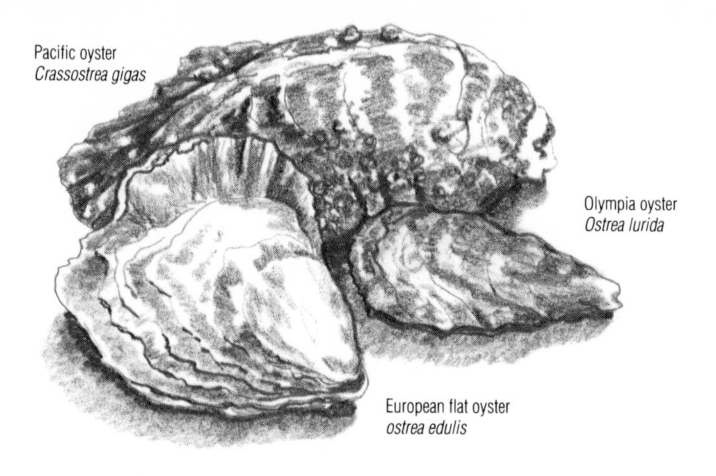


Three oyster species are cultivated in Puget Sound. The Pacific oyster (top) was introduced from Japan. It is elongated and may grow to 12 inches in length, but is usually grown to 4 to 6 inches long before harvesting. The Olympia oyster, native to this region (bottom right), is usually less than 2 inches in length. The European flat oyster (bottom left) has an almost circular shell, and grows to 4 inches in length.

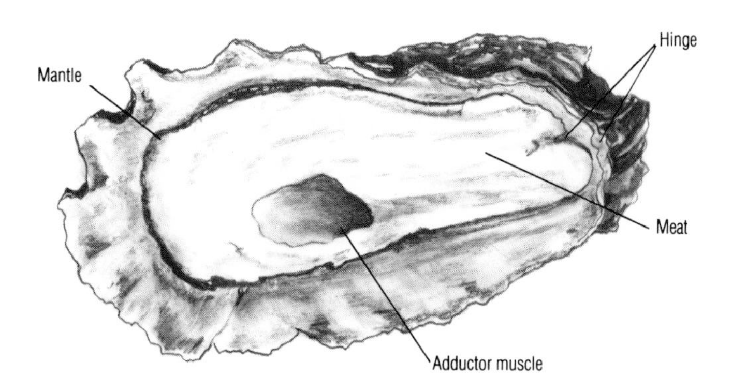


A half-shell Pacific oyster, with the right shell removed to show the living portion of the animal, the oyster "meat." Meat in a good quality oyster should have a plump, creamy appearance.

expensive ($1,000 to $3,000 per acre in the early 1900s), time consuming, and could not have been accomplished without cheap labor and private ownership of the land. The labor problem was solved by an available pool of low-paid Oriental workers who settled here after working on the railroads. The ownership question was resolved when Washington became a state in 1899 and the Legislature passed the Callow and Bush Acts authorizing sale of tidelands, thereby transferring control and management of oyster lands to private citizens. Highly productive oyster lands became extremely valuable; during the heyday of the Olympia oyster industry in the late 1800s, tidelands in Oyster Bay reportedly sold for as much as $20,000 per acre.

Private oyster culture was further encouraged when consolidation of land holdings was allowed. In 1915 the Legislature authorized purchase of the irregularly shaped pockets of tideland between the submerged private oyster beds and the beach, and then in 1927 authorized deeding of adjacent tidelands outright to owners of Callow Act lands.

Although the Olympia oyster industry was flourishing at the beginning of the twentieth century, demand for larger oysters—as was evident by the success of the Eastern oyster in California—did not go unnoticed by local oystermen. They tried to transplant Eastern oysters but the species failed to grow in Northwest waters. Then, in 1905, Japanese oystermen from Samish Bay imported mature Pacific oysters from Japan, hoping to establish the species through natural reproduction. The first effort failed, but Puget Sound oystermen continued their attempts to transplant Pacific oysters. In 1907 and 1908, three-year-old oysters from Hiroshima were planted in Liberty Bay near Poulsbo with partial success, and in 1912 more adult oysters were transplanted and survived.

In the years that followed, more and more young Pacific oysters were shipped to Puget Sound to grow and reproduce on local tidelands. But though the oysters could be fattened to commercial size in Puget Sound waters, their numbers did not increase through natural reproduction as quickly as the oystermen would have liked. Then, in 1919, something happened that altered the oystermen's approach to populating their tidelands with the new species: a large number of oysters destined for Samish Bay died en route from Japan, but young seed oysters attached to the adult shells (called *spat*) survived and grew quickly. The successful growth of the spat encouraged the oystermen to transplant seed oysters instead of adults because they were cheaper to buy and transport and had a better survival rate through shipping and transplanting. In addition, by using imported spat to grow Pacific oysters the oystermen no longer needed to rely on natural reproduction, which had been unpredictable and disappointing.

The first commercial shipment of oyster seed from Japan arrived in

Puget Sound in 1921. Dr. Trevor Kincaid, a young zoology professor at the University of Washington, recognized the value of this rapidly growing shellfish, and in the late 1920s promoted the introduction of Pacific oysters into Willapa Bay as well.

Pacific oyster culture expanded rapidly because of the Japanese seed: by 1935, West Coast seed imports rose to over 71,000 two-bushel cases of mother shells, with 12,000 to 40,000 spat per case (this case unit is still used by oyster growers). During the Depression of the 1930s the Pacific oyster industry expanded rapidly and by 1941 production in the state exceeded 9.6 million pounds, the bulk of it from Willapa Bay. After a decline during World War II, production again picked up, reaching a peak of over 13 million pounds in 1946.

As propagation of Pacific oysters increased, the Olympia oyster industry declined. Higher production costs, dedication of tidelands to other stocks, and increasing pollution took their toll on production of the Olympias, and they are now grown commercially by only a few growers in south Puget Sound. The success of the Pacific oyster, however, sustained an important industry and a way of life.

## Cultivating Oysters in Puget Sound

### Collecting Seed

The modern Puget Sound oyster farm is very different from the early wild stock "mining" ventures, or the present-day public seabed harvest along the Atlantic and Gulf coasts. In Washington State, oysters are cultivated on privately owned or leased intertidal waters seeded with spat obtained from natural spawning in Puget Sound, or set from larvae grown in local hatcheries.

From the 1920s until the late 1970s most Pacific oyster seed used by Puget Sound oystermen was imported from Japan, except from 1936 to 1947, because of successful natural reproduction and the disruption of imports caused by World War II. By the early 1970s, escalating costs for transportation of imported seed and tidelands lease fees for natural

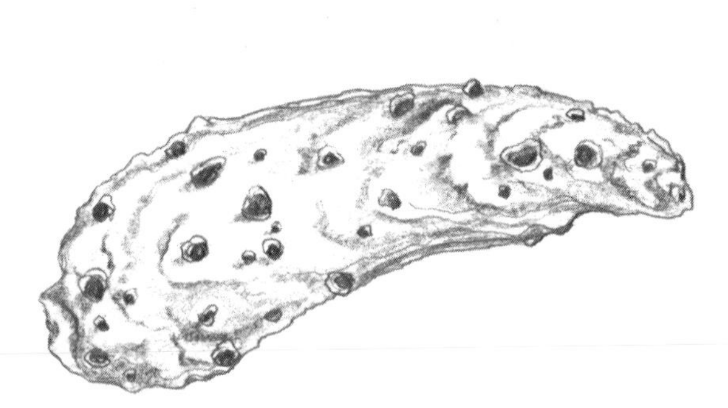

Pacific oyster spat that have set on a mother shell or cultch. These young oysters reach a size of a half inch in diameter in two months. The spat remain attached to the shell until harvest, unless they are killed by predators or disease, or are separated to reduce overcrowding.

spawning grounds motivated Washington growers to intensify efforts to establish local seed hatcheries. Most seed now used by Puget Sound oyster growers is Pacific oyster seed supplied year-round by hatcheries in Washington, Oregon, and California. Few oysters are grown from imported or wild-caught seed. It is interesting to note that Puget Sound oyster growers are a step ahead of their Japanese counterparts, who still obtain their seed exclusively from wild stock sets.

Experimental hatchery spawning of oysters was first attempted in the mid-1800s, but the technology to successfully mass produce seed was not fully realized until the mid-1960s. Modern commercial oyster hatcheries are able to produce and sell free-swimming eyed larvae that are ready for setting, and two types of seed oysters, either attached (*cultched*) or not attached (*cultchless*) to mother shell. Most growers use larvae or cultched seed: cultchless seed is primarily intended to grow single oysters for the half-shell market, which still represents only a small portion of all the oysters produced.

Larvae are produced in a hatchery under controlled conditions. The grower provides the shell cultch, setting container, and warmed seawater. The larvae are added to the water in the setting tank and setting on the cultch usually takes place within several hours. Smaller growers who don't have their own hatchery capability purchase larvae from the commercial hatcheries and set the larvae themselves.

The ability of hatcheries to supply eyed larvae has especially benefitted the industry. Unlike seed attached to bulky mother shells, larvae can be shipped by the millions in a gallon jar held in a small ice chest, which minimizes transportation costs and also eliminates introduction of disease or parasites. Another major advantage of larvae is the lack of restrictions on interstate shipment: because they are not recognized as carriers of predators or diseases (which mother shell is), they aren't subject to state inspection or quarantine.

Hatcheries have eliminated uncertainties of supply and seed quality, and although the initial expense is higher than catching wild seed (about $20 per case in 1983), predictability of supply and ease of transport make hatchery seed very cost effective. Large numbers of larvae are needed to accommodate the demands of the oyster growers. Over 16 billion larvae were purchased and set by Washington growers in 1985. Before local hatcheries began producing Pacific oyster seed, many growers collected seed from wild stocks and some still do. Puget Sound's most productive natural spawning ground is in northern Hood Canal, where good setting occurs about seven years out of ten. Hood Canal produced an equivalent of 50,000 cases of seed in 1977 and again in 1978. Spawning in other areas of the Sound is infrequent and is usually not sufficient to ensure the spat density required for sets used in commercial production.

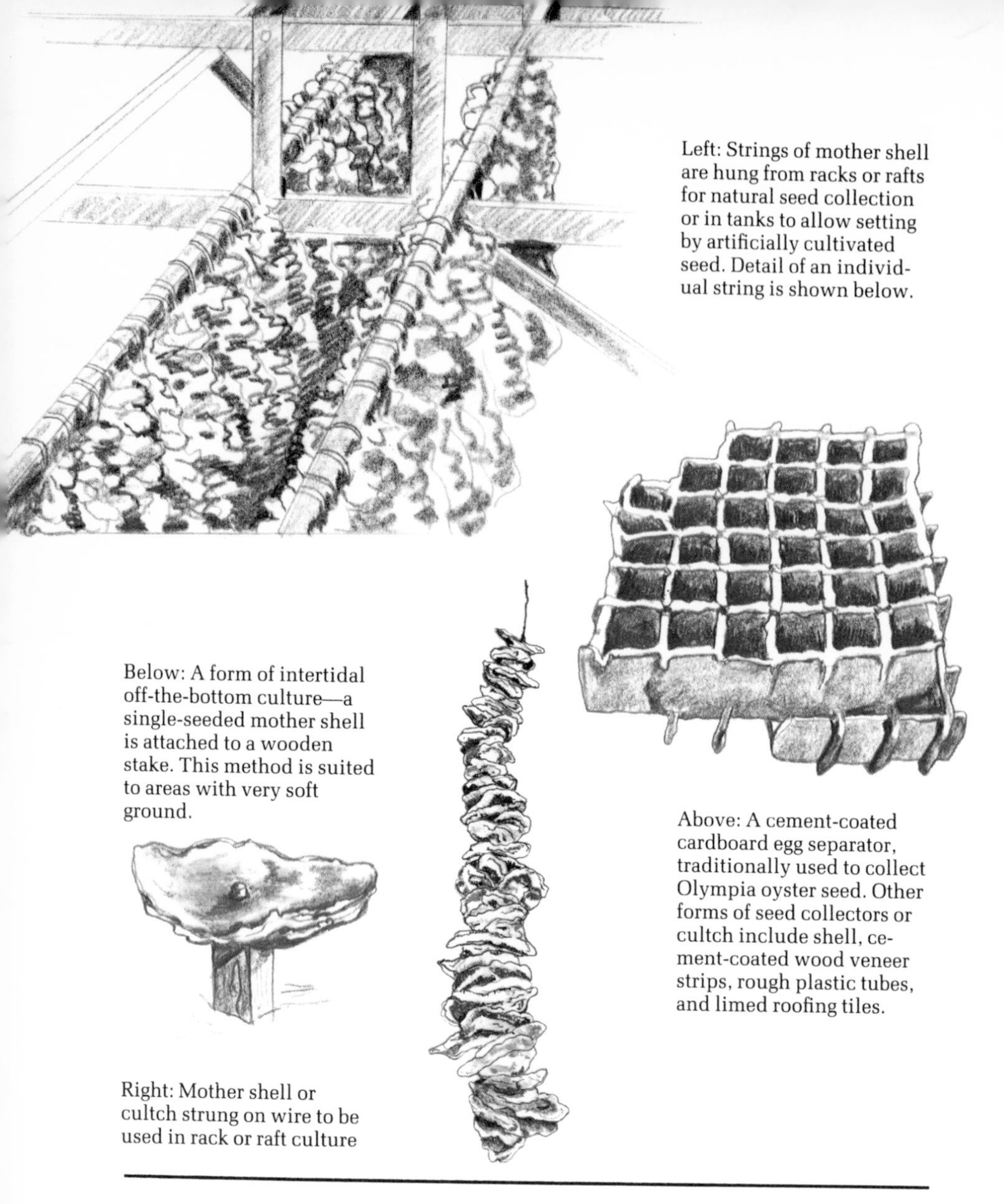

Left: Strings of mother shell are hung from racks or rafts for natural seed collection or in tanks to allow setting by artificially cultivated seed. Detail of an individual string is shown below.

Above: A cement-coated cardboard egg separator, traditionally used to collect Olympia oyster seed. Other forms of seed collectors or cultch include shell, cement-coated wood veneer strips, rough plastic tubes, and limed roofing tiles.

Below: A form of intertidal off-the-bottom culture—a single-seeded mother shell is attached to a wooden stake. This method is suited to areas with very soft ground.

Right: Mother shell or cultch strung on wire to be used in rack or raft culture

The most common method used to collect wild Pacific oyster seed is to suspend from rafts, or set on racks, mother shell that is loosely bundled in plastic mesh bags or strung on wires. The rafts or racks are placed in waters where the best sets are anticipated—usually in Dabob and Quilcene bays—from the beginning of the setting season in July and August until the seed is large enough or "hard" enough (firmly cemented to the mother shell) to withstand being moved back to the culture beds. Seed is also collected on mother shells spread on the bottom of shallow bays. This technique is employed only when very heavy set-

The interior of the Coast Oyster Company oyster hatchery located at Quilcene, Washington. These large tanks contain single-celled algae, which are fed to developing oyster larvae.

ting densities are expected, however, because raft methods have the advantage that spat are more uniformly distributed on mother shell, which facilitates later planting.

Although limited quantities of Olympia oyster seed are becoming available from hatcheries, Olympia oyster growers generally rely on natural sets that occur on their own tidelands. They catch seed by laying cement-coated egg crates, plastic pipe, or mother shell directly on the substrate or on racks. In the early 1900s the state established oyster reserves in Willapa Bay, Oakland Bay, Eld Inlet, and Totten Inlet to ensure the continued existence of Olympia oyster spawning grounds. The reserves still exist, but they no longer contain populations of Olympia oysters large enough to supply commercial quantities of native seed.

Collecting seed from wild stock is less costly than importing or buying it from hatcheries, but good spatfalls are unpredictable. Timing and intensity of spawning are affected by environmental factors (such as temperature, salinity, and exposure during low tide), by the condition of individual adult oysters, and by the concentration of spawning oysters in an area. Although domestically caught seed will always remain an important source of spat, rising costs of shipping cultch to and from the setting areas, mother shell quarantines, and unreliability of obtaining a commercially valuable set will continue to place hatcheries in a dominant role in seed production.

Planting of seed on tidelands changed the face of oystering in Puget Sound from a hunting and gathering endeavor relying solely on natural reproduction and standing stock abundance to a managed crop. Oystermen can anticipate the oyster demand for several years hence and adjust production by varying the amount of seed purchased, acres planted, and oysters harvested. The growth of hatcheries advanced the industry by ensuring reliable seed supplies and laying the groundwork for genetic manipulation by maintaining successive generations to develop strains of oysters with improved shape, size, and taste.

**Ground Culture and Harvest**

The most common method of growing oysters in Puget Sound is ground culture: mother shells—either whole or broken into smaller pieces—with seed attached are spread directly onto the oyster beds and left to grow. Pacific oysters are either planted directly on commercial grow-out beds (at 10 to 100 bushels per acre, or an average of 60 seed oysters per square yard) or on seed beds, generally areas with lower salinity where there is less likely to be predation by oyster drills and crabs. It may take one to three months for spat on seed beds to reach 10 millimeters, the minimum size at which they can be expected to survive predation. At that time they are transferred to grow-out beds to mature to market size. If there is insufficient natural food for the oysters in an area, the grower may move them to fattening beds in richer waters for six months to a year.

After one or two growing seasons, clusters of young oysters develop as those seed attached to the same mother shell grow to two or three inches in diameter. The clusters are broken up by hand or with a harrow and spread more evenly over the bed. Oysters may also become *reefed*, crowded together by storm or current action, and must be spread out, again by hand.

Although ground culture methodology is simple for Pacific oysters, the Olympia oyster requires a more controlled environment. The dike culture that proliferated in southern Puget Sound in the early 1900s is still the primary method used to cultivate Olympia oysters in south Puget Sound. The dikes are concrete or wood structures, one- to two-feet high, and enclose about one acre. Inside the dike, the bottom is leveled and gravel added to provide a firm substrate for the young seed oysters. It is a very labor-intensive process: seed must be collected and planted by hand, dike structures maintained, and competitors for food

Hand harvest on a Pacific oyster bed. Workers using forks and shovels carry the oysters to a small scow, where dead shells and undersized oysters are culled from the marketable crop.

(such as the slipper shell limpet) and predators weeded out. Some growers hold their Olympias in flat plastic mesh bags, thereby reducing some of the cultivation costs. However, Olympia oysters are still much more expensive to cultivate than Pacifics because of the higher labor and infrastructure costs and the lower yield of the small oysters.

At harvest, market-sized oysters are selected and the rest are returned to the grounds. Oysters grown on the ground are picked from the beds by hand when the tide is low, or by mechanical harvesting when it is high. Although large operators can afford to use large power harvesters—one local oysterman even has a specially designed airlift harvester—small growers usually harvest their oysters by raking or shovelling them into scows at low tide and then floating the load to shore at high tide for processing. Where individual beds are small or irregularly shaped, oysters must be harvested by hand.

Pacific oysters reach marketable size (130 to 150 per gallon) two to four years after planting. An acre of Pacific oysters yields on the average 600 to 800 gallons (a heavy crop may yield up to 1,000 gallons) of shucked meat, or about 7,800 pounds of unshucked oysters. Average annual production of Olympia oysters is about 150 gallons per acre.

Ground culture is the most economical method of growing oysters,

A mechanical oyster harvester developed by the Olympia Oyster Company in Oyster Bay. This machine uses a combination of water jets and an air lift (see insert). It is pulled along the bottom by a self-propelled barge that also holds the machinery to supply compressed air and water to the harvester.

but it does have some drawbacks. Spat on the underside of the mother shell—sometimes even the entire shell—can be smothered by silt. Predators such as crabs, starfish, and oyster drills can attack and destroy small oysters, and must be removed by hand at low tide. In densely planted areas, oysters in the middle of the bed may not fatten because those surrounding them strain the food organisms from the water first. Despite these problems, ground culture has been and will likely continue to be the predominant method of cultivating oysters in Puget Sound.

**Off-bottom Culture Methods**

There are several methods used to grow oysters by suspending them off the bottom, all of which are suitable for Puget Sound. *Rack* and *raft* cultures, developed by the Japanese, are based on fences or racks built in the intertidal zone or rafts anchored over deeper waters. Mother shell with spat are strung on wires, suspended from a horizontal pipe or timber, and hung above the bottom. A similar method is to nail mother shells to wood stakes, one to three feet above the bottom where the feeding conditions are best. Finally, in the intertidal zone oysters are sometimes grown off the ground using trays or plastic mesh bags filled with oysters on metal racks. A similar technique is to suspend *lantern nets*, basket-shaped enclosures, from floats or fixed structures. Trays and net bags have the advantage of being suitable for shallow water areas and able to accommodate both cultched and cultchless seed.

Off-bottom culture procedures are especially advantageous during the oysters' early growth stages, when they are susceptible to siltation and ground-dwelling predators such as oyster drills and flatworms. (An exception is oysters attached to stakes driven into the substrate; they are protected from siltation, but not from predation by starfish and oyster drills, which can crawl up the stakes.) In addition, unlike ground cultured oysters, which are out of the water during low tide, oysters suspended from rafts or floats are always immersed and can feed continually and thus grow faster. The suspended methods are especially well suited for grounds that are either too soft for normal ground culture (oysters sink in soft mud), or where the water is too deep to allow effective harvest.

More extensive use of rack and raft culture has the potential to increase oyster productivity in Puget Sound. In Japan, over 95 percent of oysters produced are grown on rafts, which annually produce 52,000 pounds per acre; a total annual crop of 275,000 tons of whole oysters. Despite impressive Japanese production figures, raft culture in Puget Sound is principally limited to a few rafts being used to catch seed stock, and some lantern nets used by one San Juan Island grower to

Above: Plastic mesh bags are supported on racks in shallow waters and used for the growout of individual (nonclumped) oysters destined for the half-shell market. These racks are located at Little Skookum Inlet.

Left: Detail of mesh bags shown above, attached to PVC pipe racks placed in the intertidal zone at Race Lagoon. The use of such bags has greatly reduced the effort required to handle the oysters.

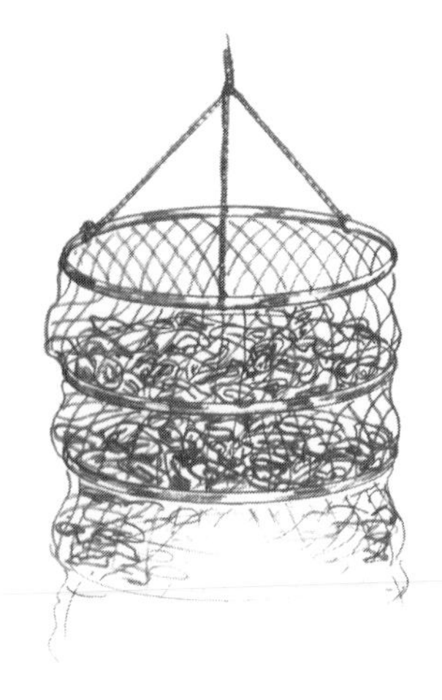

Right: A lantern net, a multilevel net tray for suspended cultivation of oysters and other shellfish. These nets are about six feet long and are well suited to culture of individual oysters. A lantern net operation is located near San Juan Island.

produce oysters for the half-shell trade.

Several growers in south Puget Sound and in Willapa Bay and Grays Harbor on Washington's Pacific Coast are now successfully using rack culture methods to rear Pacific, Olympia, and European flat (*Belon*) oysters. Raft culture is well-suited to the protected bays and estuaries of Puget Sound and is used for seed collection, but there are a number of reasons why it is not used more extensively. Puget Sound contains unused intertidal oyster beds suitable for ground culture, which is less costly to develop than building racks or rafts with equivalent rearing capacity. It should be noted, however, that the additional investment required for suspended culture is offset by production of uniformly sized and shaped oysters, which fetch the higher prices paid in the half-shell trade. Another serious problem of raft culture in certain locations of the Sound is fouling by marine organisms (usually mussels), which slows oyster growth. Finally, raft culture occurs in navigable waters and must be approved by a number of government agencies in a complex, time-consuming, review-and-permit process; it is also considered unsightly by some landowners, who cherish their view of a Puget Sound uncluttered by aquaculture paraphernalia.

## Oyster Growth and the Environment

Pacific oysters are amazingly hardy, but they are not completely immune to environmental changes caused by natural phenomena and human intervention. Abrupt changes in water temperature or salinity can kill young oysters. Strong waves or currents in the intertidal zone can wash away or bury small seed oysters, and heavy deposits of silt can smother them. Heavy freshwater runoff from storm drains and rivers can lower the salinity of estuarine waters and dump large loads of sediment and pollutants onto tidelands. Wave action from storms can suspend sediment from the substrate and tumble oysters across the bottom, accumulating them in reefs. Filling or channel dredging of nearby bottom lands can alter currents and wave patterns, thereby disrupting distribution of nutrients and other suspended materials, including oyster larvae.

Less dramatic changes, such as seasonal fluctuations in temperature and salinity, are also critical to oyster growth. Because oysters are cold-blooded animals and can cease feeding during the winter, feeding rate—and thus growth—depends on water temperatures. In addition, oysters are euryhaline animals (adapted to a wide range of salinity), and although survival is generally not affected by salinity, meat quality is. Oysters from low-salinity waters have high-yield meat but a bland flavor, and there is high water loss after shucking. In high salinity areas, the meat is salty and flavorful, but meat yield is lower.

**Food Requirements**

An oyster grower cannot simply select a site and supply his crop with the best food. Quantity and quality of natural food supplies vary from area to area, from season to season, and even between tides. Oysters feed on a variety of water-suspended materials ranging from bacteria, protozoa, small single-celled plants, larvae, and other detrital material. The best oyster lands, those that will produce "fat" marketable oysters in the shortest possible time, are where the water is of intermediate salinity and warm enough to stimulate feeding activity, where nutrient levels are high enough to ensure adequate food generated by phytoplankton, and where there are few predators and disease-causing bacteria.

Oyster growers in the Sound have not generally had their crops destroyed by any major devastating diseases. In Puget Sound, annual mortality of oysters due to disease and other factors varies from less than 4 percent to about 50 percent. The lower figure is considered normal; the high rate is an indication of extraordinarily high losses. Generally, the highest mortalities occur during unusually warm summers, at the upper end of estuaries where waters are warm, turbid, high in nutrients, and low in salinity. The factors common to summertime mortalities are a prolonged period of partial or no spawning, and decreased animal starch (glycogen) levels or reduced energy reserves. Some growers believed that such mortalities were associated with Japanese seed, because as the use of domestic seed stock increased, mortalities declined. (Studies conducted at the University of Washington have indicated that mortality-resistant stocks may be developed through selective breeding of disease-resistant genetic stock.)

**Predators**

Crabs, starfish, flatworms, Japanese oyster drills, and some species of fish prey upon young oysters, but only starfish are serious predators on adults. Predation is primarily a problem with ground-cultured oysters. Starfish inhabit the lower intertidal and subtidal zones and have to be removed by hand. The oyster drill (*Ocenebra japonica*, actually a small marine snail), which was introduced from Japan with imported seed, is another serious predator, particularly during the first year. Rigid inspections have nearly eliminated oyster drills from imported seed, and intrastate regulations governing the transport of shell help prevent transfer of drills into uninfested areas. Predation by oyster drills and crabs is controlled by planting seed on upper intertidal beds in brackish waters or on an off-bottom apparatus, such as racks or rafts.

Because of their smaller size, Olympia oysters are more susceptible to the effects of drills, flatworms, and predation by waterfowl than the Pacifics. Drill and flatworm predation is controlled to a certain extent by maintaining the oysters in diked, low salinity waters.

Other organisms such as slipper shells, barnacles, mussels, and sponges compete with oysters for space and food. Modifications in culture procedures have managed to eliminate the effects of these competitors, and they are no longer considered to be factors that limit oyster farming.

Ghost shrimps of the genera *Upogebia* and *Callianassa* are obnoxious pests that create problems in ground culture areas. They burrow into the oyster beds and loosen the substrate, which smothers the seed and causes adult oysters to sink into the sediment. Oystermen cope with these pests by avoiding infested beds or treating the grounds with *Sevin*, a pesticide. Oystermen would prefer a nonchemical alternative to Sevin, but there is no simple way to physically extract ghost shrimps from soft sediment. Because of the unique nature of this pest control method (unusual in that it is applied to a marine organism), the Washington State Department of Fisheries has reviewed its probable environmental effects. The department concluded that when beds were treated correctly there was little short- or long-term carryover of the killing effect of Sevin to adjacent waters or seabed, and that the benefits to the growers outweighed the possible negative consequences.

---

Right: The ghost shrimp, a burrowing shrimp-like crustacean, up to six inches in length. Very dense populations of ghost shrimps and mud shrimps can cause serious harm to oyster ground by making it too soft for oyster culture.

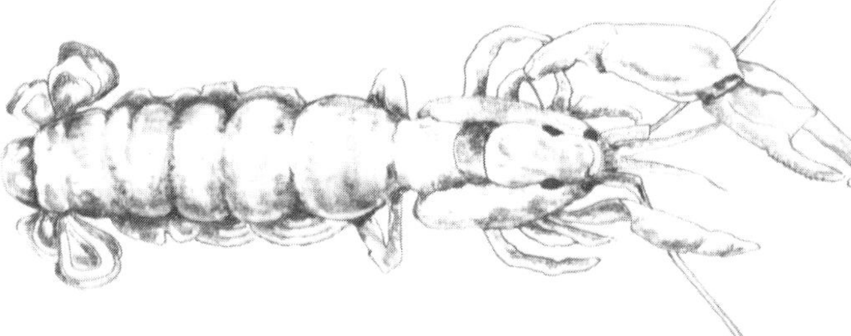

Left: The oyster drill, shown here with a cluster of eggs, is one of several species of marine snails that prey on oysters by rasping a small hole through the shell and using an extendible mouth part to consume the meat.

**Problems with Pollution**

Because the oyster is an efficient feeder, over time it can accumulate pollutants from the environment, which can slow its growth, impair breeding activity, and even cause death. Even if the oyster itself is not affected, pollutants concentrated on an oyster bed may indicate a public health hazard serious enough to prevent commercial marketing of the crop. Domestic wastes from city sewer outfalls and leachates from septic tanks can contain high concentrations of nutrients, bacteria, viruses, oil and grease, heavy metals, and pesticide residues. Even treated waste from sewage treatment plants carries some pollutants into the Sound, and treatment plants may be inundated by storm runoff during heavy rains, or work ineffectively when sewage volumes are low.

Water that has been polluted by human sewage is hazardous because several diseases, such as typhoid, cholera, dysentery, and hepatitis, are transmitted by this route. *Indicator organisms* of these diseases (so-called because they are supposed to indicate the presence of vertebrate fecal matter) are measured and correlated with the number of pathogens in a water sample. The standards are strict: at least half the water samples taken over shellfish beds must have fecal coliform levels equal to or less than 14 MPN per 100 milliliters of water. (MPN, "Most Probable Number," is a statistical measure of bacteria found in the water.) Growing areas are carefully monitored by the Washington Department of Social and Health Services (DSHS), to guarantee the safety of shellfish to the consumer.

Further sources of pollutants are recreational and commercial boating activity (particularly near marinas and heavily used anchorages) and storm runoff. Storm runoff contains bacteria and other pollutants that are carried to oyster grounds via streams and overland flow. Most of the bacteria are derived from the soil; however, indicator organisms derived from feces from cows, chickens, dogs, cats, and wild animals can also occur in land runoff. The level of these indicator organisms can be high at all times, even when there may be no obvious source of pollutants. Unfortunately, there is no clear answer to the question of health risk of bacteria in stormwater. Wild and domestic animals usually carry no significant waterborne human pathogens, and the moderate levels of coliform bacteria in stormwater probably present little direct threat to human health.

Exposure to industrial wastes such as spent sulfite liquor (SWL) from pulp mills may kill oyster larvae and cause slow growth, poor meat condition, and death in adult oysters. The decline of oyster populations—especially Olympia oysters—near Shelton in Oakland Bay, in Samish Bay, near Bellingham, in Port Susan, and south of Everett was attributed by oyster growers to spent sulfite waste liquors from pulp mills. That oysters may be harmed by the sulfite liquor was supported

Table 2.1 Areas within the state of Washington where commercial shellfish harvest has been closed or restricted due to bacterial contamination. These areas are commonly associated with urban centers, marinas, and sewage outfalls.

| Date | Location | Cause |
|---|---|---|
| **Decertified** | | |
| 1950s | Dyes Inlet—all | Bremerton STP* |
| 1950s | Sinclair Inlet—all | Bremerton STP |
| 1950s** | Oakland Bay/Hammersly Inlet—in the vicinity of Shelton | Shelton STP, Mill |
| 1950s | Budd Inlet | STP, Deschutes River |
| 1960s | Liberty Bay—east side, near Poulsbo | Poulsbo STP, Marina |
| 1960s | Grays Harbor—about 1/4, basically the eastern lobe | STPs, Mills |
| 1960s | Willapa Bay—about 10%, around mouth of the Willapa River | Raymond, South Bend STPs |
| 1981 | Port Susan—about 1/3 of the tideflats near the Stillaguamish River | Dairy runoff into the Stillaguamish River, STPs |
| 1981 | Burley Lagoon—all | Nonpoint sources |
| 1982 | Minter Bay—all | Nonpoint sources |
| 1985 | Quilcene Bay—around streams at head of bay | Nonpoint sources |
| 1985 | Henderson Inlet—lower quarter of inlet | Nonpoint sources |
| **Conditionally Approved** | | |
| 1982 | Eld Inlet—Lower quarter of inlet | Nonpoint sources |
| 1983 | Penn Cove—portion of north shore | STP or nonpoint sources |
| **Uncertifiable***** | | |
| | Eastshore of Puget Sound from Tacoma to Edmonds | STPs, industrial |
| | Hartstene Island, northern tip | Private STP |
| | Port Townsend | STP |
| | Winslow | STP |
| | Appletree Cove, near Kingston | Sewage Outfall |
| | Port Gamble | Sewage Outfall |
| | Everett | STP, industry, nonpoint sources |
| | Bellingham Bay | STP, mills, nonpoint sources |

* Sewage Treatment Plant

** Decertified area reduced in 1980 due to installation of secondary treatment.

*** Based on review of geoduck beds for lease suitability by the Washington Department of Fisheries and DSHS.

Source: Department of Ecology, 1984. "Shellfish Protection Strategy."

by laboratory testing. Only the sulfite process was implicated in oyster kills. Kraft pulping process effluent was found to be less toxic to shellfish and can be adequately treated by standard sewage systems. A great deal of effort was spent cleaning up sulfite mill wastes, with significant improvements noted around Everett and Port Angeles. There are no longer any pulp mills operating in southern Puget Sound.

Despite the existence of pollutants, careful environmental inspection ensures that commercially harvested shellfish are safe for human consumption. Growing areas are classified as follows:

> *Approved for commercial shellfish harvesting*— oysters may be cultured and harvested year-round;
>
> *Conditionally approved*—occasional closures result from increases of bacteria associated with freshwater runoff; when rainfall exceeds an upper limit (presently 1½ inches in 24 hours), the beds are closed for two weeks;
>
> *Closed to commercial shellfish harvesting*—decertified and uncertifiable waters that would produce shellfish unsafe for human consumption.

Shellfish may be taken from closed waters only for personal use, but recreational harvesters should identify the cause of the closure and should not eat the oysters raw. The Federal Food and Drug Administration oversees the state shellfish inspection program, and the state administers it and carries out the actual inspections. All oyster growing areas in the state of Washington are classified as Approved for Commercial Shellfish Harvesting, with the exception of those areas listed in Table 2.1.

It is possible to treat oysters that have unacceptable concentrations of fecal bacteria by moving them to uncontaminated grounds or holding them in special treatment systems, a process called *depuration*. It adds to production costs, however, and is currently not considered to be a permissible alternative by DSHS.

Most shellfish farmers believe that consideration must be given to coastal zone planning at the local level to ensure provision and maintenance of suitable areas for shellfish growing and harvesting. Also, the public health significance of various pollution indicators emphasizing coliform bacteria needs to be more specifically defined. As was noted above, this widely used indicator method does not necessarily reflect contamination by human waste but can be due to agricultural and urban runoff or even local wildfowl populations. There is a need to provide adequate public health protection without restricting the utilization of oysters and other shellfish that are, in fact, safe to eat.

# Commercial Oyster Production

### Production Data

World oyster production in 1981 (all types) was about one million tons (whole weight). The U.S. production was nearly 350,000 tons—consisting of 320,000 tons of Eastern and 30,000 tons of Pacific oysters. For comparison, Pacific oyster production in Japan in 1981 was 235,000 tons.

Production of Pacific oysters in Puget Sound has not varied greatly since 1935, and now accounts for about 30 percent of the total West Coast production. Washington was responsible for 77 percent of the Western Region (states of Washington, California, Oregon, and Hawaii, and the Pacific Territories) gross market value, GMV, generated by sales of adult oysters in 1981. Nearly half of all single oysters, 81 percent of the shucked oysters, and much of the seed are produced here. The industry is characterized by five large farms, which each sell about 100,000 gallons (875,000 pounds) of shucked meat annually, and numerous small farms with annual sales of less than 20,000 gallons (175,000 pounds) each. Approximately 29,000 acres are privately owned or leased from the state, however only 3,200 were harvested in 1981.

Pacific oyster production in Puget Sound in 1982 was nearly 2.3 million pounds of meat, with a processed value of over $7.7 million. Total production in Washington State during the same period was over 6 million pounds with a processed value of over $20.4 million. In 1982 Washington State oyster production amounted to 11 percent of total U.S. production, three percent of total state fisheries catch, and ten percent of processed value for all state fisheries products. By comparison, production in Oregon, California and British Columbia each averaged less than one million pounds per year for the same period.

Olympia oyster harvests peaked in 1926 at 701,000 pounds, and decreased to about 2,700 pounds in 1982. The decline in South Sound production from the late 1920s through the 1950s was attributed to increased mortalities caused by pulp mill effluent in Oakland Bay and Japanese oyster drills, and to reduced harvesting because of depressed prices and increased labor costs. Production increased slightly after the pulp mill in Shelton closed in 1957, but declined again after the early 1970s.

Historically, annual oyster production in Washington State peaked at over 13 million pounds in 1946; it has since decreased to about half that amount. It is interesting to note that during this same period, domestic landings of oysters in the rest of the United States also declined because of overharvesting, natural disasters, disease, and pollution. In

Total oyster production (pounds of shucked meats) in the state of Washington (including Puget Sound, Willapa Bay, and Grays Harbor) compared with Puget Sound Pacific oyster and Olympia oyster production. (Source: Annual landings data for 1935—82, Washington Department of Fisheries)

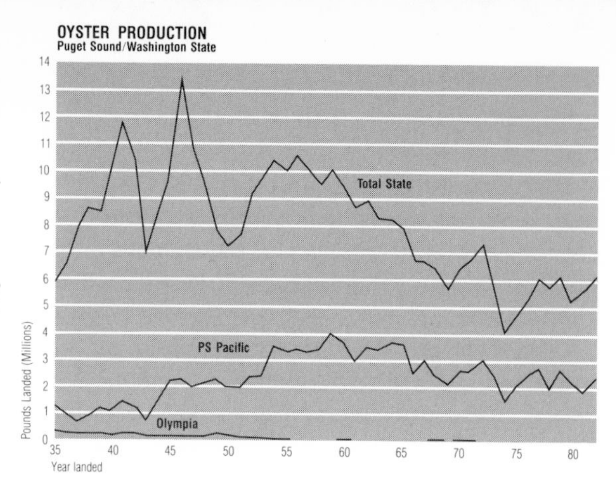


Puget Sound Olympia oyster production (pounds of shucked meats). Production in 1982 was 2,674 pounds. (Source: Annual landings for 1935–82, Washington Department of Fisheries)

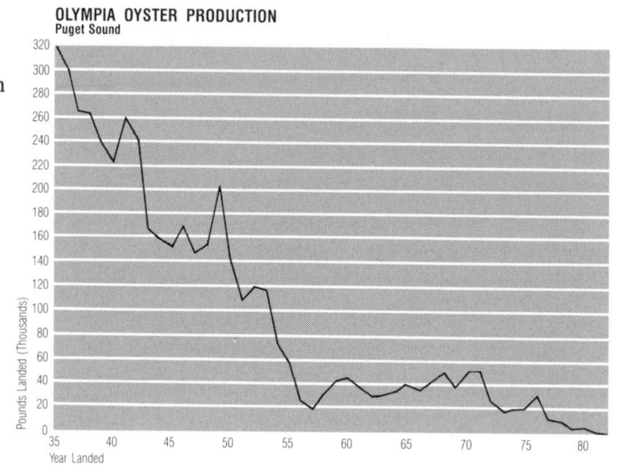


Total Puget Sound Pacific oyster production, shown along with north, central, and south Sound production (pounds of shucked meats). (Source: Annual landings data for 1951—82, Washington Department of Fisheries)

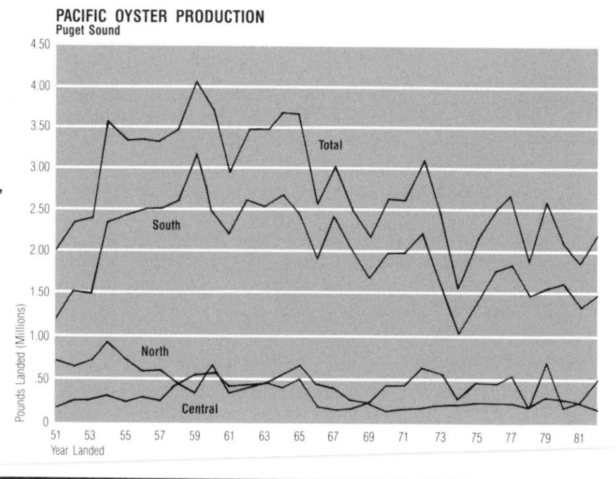

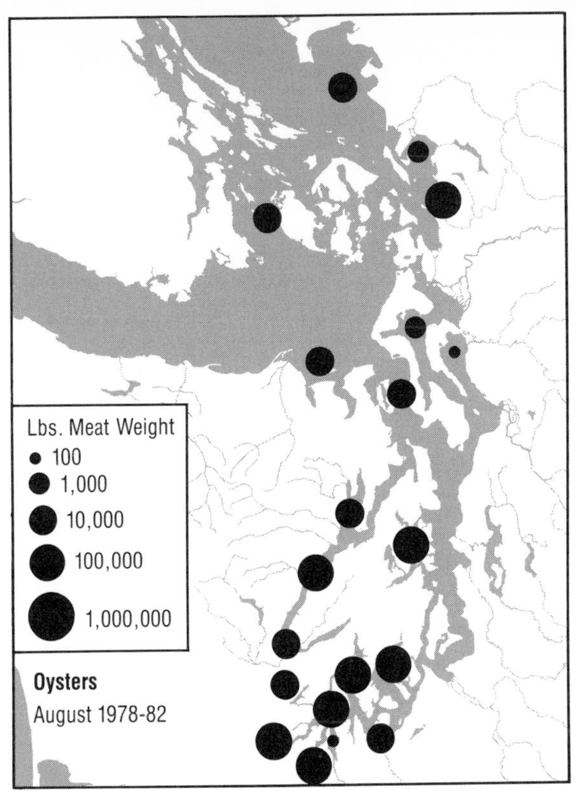


Distribution of oyster landings in Puget Sound, 1978–82. Each circle depicts average annual landings from the value given up to the next highest value.

1929, 90 million pounds (meat weight) were landed in the United States, and from 1977 through 1981 annual landings averaged 49 million pounds; the value of landings, however, increased despite the decrease in pounds harvested.

**Processing and Marketing**

Processing is the most labor-intensive phase of the oyster business because oysters must be shucked individually by hand. Most Olympias are sold in the shell; shucking is very limited. Most Pacific oysters are sold fresh-shucked to retail and institutional markets, and the rest are breaded and frozen, or canned—either boiled or smoked. Few Pacific oysters are sold in the shell, although there is an expanding market for this product, which fetches a higher price to the grower. Shucked oysters are sorted into standard market sizes ranging from cocktail oysters at 38 or more per pint to large oysters at 6–8 per pint) and packed in various sized containers (from 10 ounces to one gallon). At one time, most oysters were canned, but low-priced canned imports from Japan and Korea made canning in the United States uneconomical. This situation may be changing, however, as the cost of imports rises.

Many smaller Puget Sound growers sell whole oysters to a proces-

The Olympia Oyster Company processing plant located at Oyster Bay, near Shelton, Washington. Oyster shucking (right) in this plant has changed little since the early part of this century.

sor who handles shucking and marketing. This specialization is particularly successful when the demand is great for oysters, but can work against the small operator when supplies exceed demand. The small grower may be unable to advertise and promote his product as effectively as the large processor at these times.

The 1982 processed price for fresh, shucked Pacific oysters sold in the Puget Sound area exceeded $29 per gallon ($3.35 per pound). It is interesting to note that in 1923 the price of Pacific oysters was $14.00 per gallon, but in the 1930s they could be bought for as little as $1.07 per gallon. Times were hard for oystermen, few made money, and many growers were forced out of business.

Market demand for fresh oysters increased dramatically during the 1970s, in response to the demand for affordable, natural, unprocessed food items. In Washington, about 700,000 gallons of shucked oyster meats were sold in 1982, and single oyster sales in the same period were 443,000 dozen (equivalent to about 31,000 gallons of meats). Improvement of the oyster beds, development of new seed sources, application of intensive culture procedures, and development of genetic improvements in oyster stocks have increased production to meet increasing demand for oysters. Demand is especially high for oysters grown off-bottom or on a hard substrate as these stocks are reported to have a milder flavor than oysters raised on muddy bottoms. Considerably higher prices are demanded of the small Olympias, the Belons, and smaller Pacific half-shells. For Olympia oysters, however, supplies are much lower than they were in the early days of oyster culture in Puget Sound. Slow growth rate and intensive labor involved in culturing, harvesting, and processing discouraged the industry. As a result of high production costs, limited supplies, and lack of competition from similar products, the cost of native oysters is considerably higher than Pacific oysters. The price per gallon (at about 2,000 hand-shucked oysters, or two to three hours' shucking time per gallon) in May 1979, when they were available, was up to $120.

Much of the state production—65 percent of the processed oysters, 36 percent of the singles, and 64 percent of the seed oysters—is exported out of state. Oyster growing is labor- and capital-intensive, accounting for employment of nearly 800 people in Washington State. Major capital investment is also required: one oyster producer recently spent $3.5 million to build a new shucking and processing plant on Willapa Bay, and similar expenses have accrued to other oyster farmers.

Processing and marketing limitations have been partially responsible for lack of development of the oyster industry. Attempts to mechanize oyster shucking have not yet succeeded, neither have pasteurizing techniques that would help retain flavor and texture while enhancing shelf-life. Also lacking are advances critical to development of new processed products such as dehydrated soups. Unfortunately, the limited financial resources of the industry have hindered both technological research and promotional campaigns.

## The Future of the Oyster Industry

### Osyter Farming and Land Use in Washington

Historical use suggests that an immediate 100 percent increase in oyster production is possible without expansion of existing growing areas. About 40 percent of the growers interviewed in a 1984 survey projected production increases via installation of artificial setting tanks, use of more intensive rearing methods, and culture of hybrid oyster strains. Additional production would be possible by opening up new culture areas and reclaiming old areas not presently used. However, the relationship of oyster culture to land use is changing rapidly. Oyster farming on private grounds has historically been considered an extension of upland farming, and except for public health considerations, has not been subjected to extensive land use regulation. In recent years, efforts to modify oyster beds and to establish commercial raft culture have come under increasingly critical public and government scrutiny. Raft culture is conducted over State land, not private land, which is outside the traditional oyster grounds. It requires a shoreline development permit, public health certification, site lease agreement, and navigable waters permit. (These requirements are discussed in further detail in Chapter Three.) Negative public reaction can result in governmental permit delays and litigation, and has discouraged raft culture in Puget Sound.

Lack of suitable oyster lands is not a limiting factor for intertidal oyster culture in Puget Sound. About 80,000 acres have been surveyed and classified for shellfish growing by the State Department of Social and Health Services. Some areas have diminished in size, however, as oyster lands have been sold and used for other purposes. Nevertheless, because the Puget Sound oystermen own or lease their oyster beds, they do not have to compete with other shoreline land uses, and conflicts are most likely to involve water quality or the aesthetic values of upland owners.

**New Pollution Threats**

Closures of oyster harvest areas due to contamination by indicator bacteria have occurred at a number of locations in Washington State (Table 2.1). Other closures of beds in north Puget Sound have been associated mainly with paralytic shellfish poisoning (see Chapter Four, "Mussels"). Historically, urban growth and the resultant discharges from street runoff, sewage treatment plants, and septic tank leachate have had the most significant impact on oyster culture. As was discussed earlier, shellfish growing areas are carefully monitored to ensure oysters are grown in the highest quality waters. Grounds are usually closed to commerial harvest near marinas, heavily used boat traffic areas, sewage discharges, and urban areas. The continued spread of urbanization places a greater economic hardship on the oyster growers by diminishing harvestable areas.

Only by fortunate circumstance did the most intense urbanization of Puget Sound develop apart from the richest culture areas, thus allowing the traditional industry to escape most of the pollution problems. Still, as is shown in Table 2.1, large areas of valuable intertidal and subtidal lands are closed to harvest. The pattern of most recent closures strongly suggests that contamination from septic tanks and stormwater runoff in lightly to moderately developed areas is now the most serious potential threat to the areas used for cultivating oysters and other shellfish. For example, located on the fringes of the cities of Tacoma and Olympia in central and southern Puget Sound are four important oyster growing areas—Minter Bay, Burley Lagoon, and southern portions of

Henderson and Eld Inlets—that have been decertified or conditionally approved. These relatively rural areas have recently undergone considerable development as more people have been lured by a hope for a slower-paced rural lifestyle. This increased human influx has resulted in greater use of onsite waste disposal (i.e., septic tanks, often in poor soils), increased small-scale animal keeping (or "hobby farming"), increased development near shorelines and creeks, and more household pets.

The problem of stormwater runoff cannot be regarded lightly, because there is a great potential for widescale contamination and closure of oyster and other shellfish growing areas in Puget Sound. This was clearly illustrated in a 1984 study conducted by Thurston County in southern Puget Sound. Particular emphasis was placed on sampling streams, tributary waters, and stormwater drainages to Henderson and Eld Inlets.

Samples taken at various locations in a stream draining a large storm sewer system exhibited a significant increase in indicator bacteria levels occurring downstream from the sewer outfall. The storm sewer drained land areas subjected to uses ranging from low-density residential to industrial, and the stormwater was extremely turbid and gray with predominant oil sheen. Small culverts, retention ponds, and roadways draining into streams that were tributary to important oyster beds in Henderson Inlet also had high bacteria concentrations. Creeks and culverts in the rural Eld Inlet watershed received drainage from forested areas, pasture lands, low-density residential areas, and roadways. Fecal coliform levels in most of the creeks were low to moderate. However, a few of the streams, culverts, and drainage ditches had bacteria levels as high as those recorded in the more populated areas of Henderson Inlet.

The numerous and diffuse nature of stormwater and other so-called *nonpoint* sources of contamination makes them very difficult to control; therefore, a considerable amount of time and effort must be expended to understand and resolve the problem. Controlling contamination originating from a marine or upland source cannot, of course, be accomplished solely by the oystermen. It takes the cooperation of landowners and a long-term commitment by local, state, and federal agencies. The control of pollutants from boats, sewage outfalls, and other "point" sources can, given proper incentive for treatment, be relatively easily controlled.

Surprisingly, there is no easy and set way to reduce the bacteria in runoff. Rather, a combination of methods must be employed, including storing runoff in ponds, providing areas where the water can infiltrate or seep back into the ground, and passing ordinances regulating animal control, stream corridor preservation, and development.

**Advances in Culture Technology**

Advances in oyster culture are occurring rapidly and growers are beginning to be able to control and manipulate stocks using selective rearing and hybridization, techniques that were not feasible only a few years ago. Genetic selection studies (such as those under way at Oregon State University and the University of Washington) are aimed at developing strains that are resistant to summertime mortality, or are not as likely to enter a spawning phase (which degrades meat quality). As the result of these and other studies, strains of oysters may eventually be developed that grow faster; are more uniformly shaped; have consistently good color, flavor, and appearance; and develop characteristics that allow them to better adapt to artificial propagation. These studies may also lead to hybrid oysters, having the favorable features of two or more species or develop sterile oysters that will not degrade due to spawning.

The industry is likely to continue its dependence on the Pacific oyster. Culture of Olympia oysters is constrained by their slow growth rate, small size, low abundance, and high labor costs. It may be more economical to raise Olympias in trays and rafts than in dikes, which would also lessen losses to predators, but raft culture still faces problems in availability of seed stocks and regulatory difficulties.

Other potential candidates for commercial cultivation in the Northwest are the European flat oyster, *Ostrea edulis*, and a small variety of the Pacific known as the Kumamoto oyster, which grows faster and larger than the Olympia. Techniques have been developed for hatchery production of European oyster seed, and several growers have shown that it is adaptable to the aquatic environment in southern Puget Sound. The Kumamoto oyster (and other *gigas* varieties) was planted in the Sound in the 1960s, but no natural stocks are available and seed supplies have been extremely limited.

The introduction of these oysters is likely to improve the capability of the growers to respond to changing market demands. Both the European flat oyster and Kumamoto oyster are relatively high cost products, especially well suited to the half-shell restaurant trade. Their high value makes it possible to grow them in the more expensive off-bottom systems where closer attention can be paid to product quality control. At the same time, unless off-bottom culture is greatly expanded, supplies of these oysters will remain limited, with very few of them reaching home consumers.

CHAPTER THREE

# Clams

During the 1940s the late Seattle restauranteur Ivar Haglund popularized the "Ballad of the Early Settler," a tale of a pioneer who failed to find his fortune in the Alaska Gold Rush, and returned to Puget Sound to live a happy life surrounded by "acres of clams." Indeed, clams had long been a favorite food for the inhabitants of the Puget Sound region because they were free for the taking. Early Western Washington settlers had a maxim about the bounty to be found in the local tidelands—"When the tide is out, the table is set." Clams were widespread throughout the Sound, were accessible, and were considered public property. Anyone within reach of the Sound's beaches could dig a bucketful in short order on a low tide and provide meat for a chowder or steamed clams at suppertime.

The extensive beaches of Puget Sound supported a small, but stable, commercial bay clam fishery in the early 1900s, mostly from South Sound and Hood Canal. The industry was based primarily on the abundant native littlenecks and butter clams, which were good for canning because of their large size, high meat yield, and excellent flavor. By 1919, South Sound commercial clam fisheries supplied most Washington—and some Oregon—markets with fresh littleneck clams. Although most clams harvested in Puget Sound were and still are sold primarily to the fresh-clam market, there was a clam cannery operating in Friday Harbor in the San Juan Islands as early as 1900. The canning industry was forced to compete with canned East Coast clams and razor clams, however, and it never prospered.

Most commercially harvested clams from Puget Sound were taken on publicly owned beaches. However, after 1940 clam harvests from these areas began to decline. This was due to a variety of factors including heavy digging pressure during the Depression of the 1930s, a subsequent lack of major setting of young clams combined with overharvesting, a reduction in beach areas open to public harvest, and poor prices.

Since the 1950s technological developments have altered the clam industry from the early days of intertidal public harvest. Subtidal hardshell and geoduck stocks have been exploited, and a softshell clam fishery was initiated in Skagit Bay and Port Susan. Finally, methods have been developed to rear clams under controlled conditions, thus en-

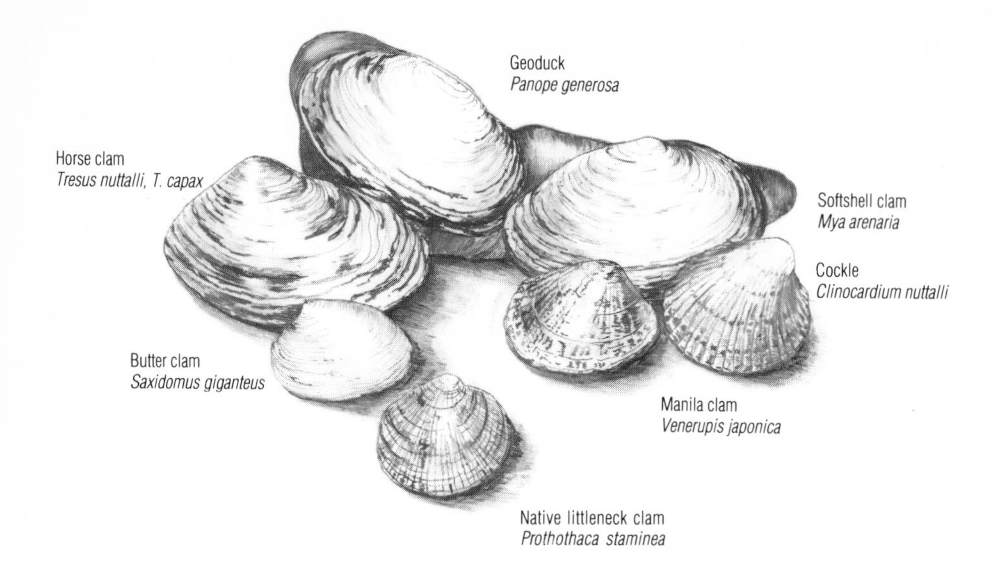


Commercially cultivated or harvested clams from Puget Sound. Clockwise from top: geoduck (shell up to 8 inches long), softshell clam (up to 7 inches), cockle (to 4 inches), Manila clam (usually to 2½ inches in Puget Sound), native littleneck (to 2½ inches), butter clam (to 6 inches), and horse clam (also called "gaper," to 8 inches).

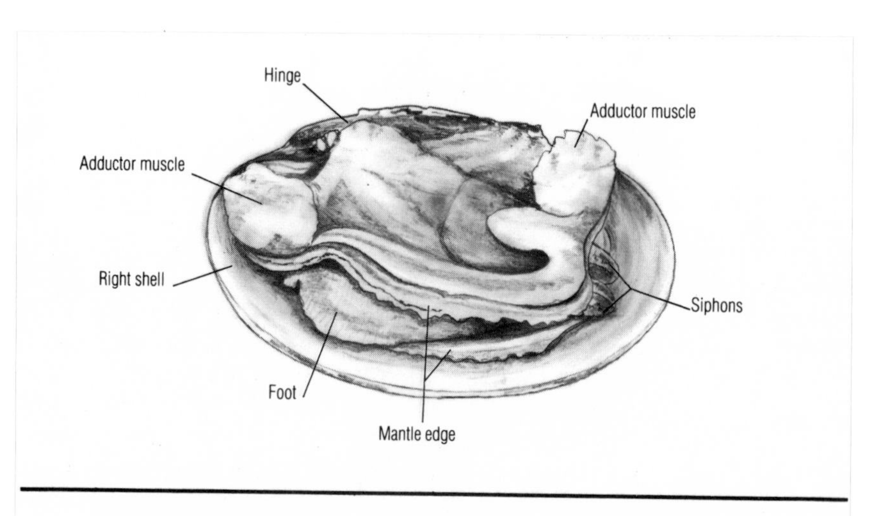


A Manila clam with one shell removed to show the living portion of the animal—the typical view when one eats a "steamer" clam. Note the two large shell closing muscles (adductors), foot, and siphons (shown retracted).

abling a clam industry to develop based on controlled aquaculture methods rather than relying on reproduction of wild stocks.

Washington is the only Western Region state (Washington, Oregon, California, Hawaii, and the Pacific Territories) where commercial quantities of clams are regularly farmed. The Manila and native littleneck clam industry is composed of about 25 aquatic farmers who own intertidal clam beds or lease tidelands from the state. Total statewide employment on clam farms numbers about 70 workers. The larger farms each harvest more than 150,000 pounds of clams annually (whole weight), while smaller farms usually harvest no more than 50,000 each per year. About half of the clam growers in the state also farm oysters, and most still depend on natural reproduction for seed.

## Growing and Harvesting Clams

### Species Harvested

Many species of clams are found along the Pacific Coast and in Puget Sound, but only eight species of bay clams are important to the commercial and recreational clam fisheries of Puget Sound. They include species called "hardshell" clams—butter clams, native littleneck clams, Manila clams, two species of horse clams, cockles, and geoducks—and the eastern softshell clams. Manila and native littleneck clam farming in the South Sound and Hood Canal is closely associated with the oyster industry, and most growers from these locales harvest both bivalves. Native littleneck and some horse and butter clams are harvested intertidally and subtidally in the protected bays of the Sound, Hood Canal, and the Strait of Juan de Fuca; softshell clams are harvested intertidally in Port Susan and Skagit Bay; and geoducks are harvested in the subtidal zone in central and northern Puget Sound.

### Regulatory Requirements

Intertidal clams are harvested from tidelands that are owned or leased from private owners or the state. Lease charges are usually calculated from a "stumpage fee," a royalty paid on the number of pounds of native stock harvested. Subtidal lands are all state owned, thus the right to harvest on these lands is granted only by the state, and hardshell clam dredgers obtain state leases for exclusive harvest. Lease tracts granted by the Department of Natural Resources must be approved by the Department of Fisheries, certified by the Department of Social and Health Services, and surveyed by the Department of Fisheries. The lessee pays to the Departments of Fisheries and Natural Resources a land lease fee, a license fee for harvesting, a royalty on the amount of public resource clams harvested, and a surveying and mapping fee.

Subtidal geoduck tracts are leased by the Department of Natural

Resources for two to four years, and are awarded to the highest bidder. In addition to lease and certification fees, the lessee must have a clam farm license, and a nozzle and harvest license. After the lease period expires, the beds revert to the state to allow renewal of the crop, which can take up to 30 to 40 years unless the bed can be replanted with hatchery cultured clams. Test plants of hatchery-reared geoducks suggest that it may be possible to reduce the rotation time of harvested tracts to as few as six to ten years.

Geoduck lease tracts are based on accessible acreage, which must be at least 600 feet seaward from the high-water line or no less than 18 feet deep, out to the 60-foot contour line, the maximum safe long-term working depth of commercial divers using pressurized air. Early leases set an allowable yearly maximum and minimum harvest for each tract as determined by dividing the tract acreage and estimated number of geoducks. Recent leases set no limits on geoduck harvest with the tract size determining total harvest. It is estimated that about five million pounds of geoduck (in the shell) can be harvested annually from all Puget Sound beds without depleting the resource.

**Reproduction and Growth Enhancement**

Clams harvested in Puget Sound grow in the substrate, and there has been no attempt to increase density by growing them off bottom. Most seed is obtained from natural stocks, rather than artificially propagated or cultured, although ground seeding methods similar to those used by oyster growers are used by the clam industry. Clam harvesters face problems of irregularity in recruitment of juveniles, difficulty in sustaining production, increased digging costs, low-density beds, and difficulty in securing harvesting permits for subtidal clams.

Habitat that clams grow in affects survival and growth. Composition of the soil, water quality, and predators can cause fluctuations in spawning intensity, and larval survival and dispersion. Approaches for enhancing the clam bed yields include:

- altering the habitat to attract natural larval setting;
- transporting naturally set juvenile clams to barren areas; and
- using hatchery-reared seed.

Transporting naturally set juvenile clams is feasible when there is an excess of juveniles, but it is site-specific and requires the grower to carefully monitor the status of his beds. Transplants have been used with some success in Japan and on the East Coast to reduce the peaks and valleys of the naturally variable setting densities. However, it is not always feasible to rely on natural settings to resupply harvested grounds because the environmental factors that make natural sets so variable also adversely affect the transplanted juveniles.

Because of the difficulty, or at least the uncertainty, of transplanting juveniles from natural stocks, enhancement of clam beds in Puget Sound has centered around hatchery-reared seed. Larvae of various species have been cultured successfully in laboratories, and techniques to grow seed clams to a size at which they survive transplant have been developed for Manila clams and geoducks. Manila clams are particularly suitable for hatchery rearing because the rearing period from spawning through setting is short, they are easy to spawn, and the incidence of disease is low. Several commercial hatcheries on the Pacific Coast now produce quantities of young Manila clams for sale.

Since 1972, University of Washington fisheries researchers' plantings have proven experimentally that hatchery seed planted in high quality beds, protected from predation, can be a viable economic alternative to obtaining naturally set seed. Factors important to cultivating Manila clams from seed stock are habitat, predation, and density and size of seed at planting. Suitable habitats are those protected from heavy wave action and composed of substrate with a high proportion of gravel in relation to fine silt. Planting on the lower intertidal zone increases submersion and feeding time, resulting in increased growth and productivity. Usually, clams planted in south Puget Sound grow fastest.

Intertidal substrate that has become less productive as a clam bed, as gravel has eroded or more sand has been deposited, can be restored by clam growers. On muddy beaches, gravel can be added to the substrate to make it more suitable for clams and protect them from predation. Improved substrate may be more conducive to natural clam setting, thereby eliminating the need to seed the area with hatchery clams. Substrate may be further stabilized by laying plastic mesh nets over the clam beds. The nets improve the recovery of clams on loose sand-gravel beaches where they might be subject to displacement by wave activity. Both net protection and substrate modification add to the cost of production and may be subject to approval by government agencies.

Experimental studies in Puget Sound and the San Juan Islands, where clams were planted at densities of from 19 to 158 seed per square foot, suggest that there is no single best planting strategy. Densely planted seed does not necessarily result in greater production because competition for food and space may retard growth, and some seed clams move to less crowded grounds. The smaller the seed planted, the higher density required to compensate for greater losses to predators and other hazards.

Timing of seed planting affects production because seed planted in the fall have higher juvenile mortalities due to wave scour (associated with strong winter winds) and low temperatures. Spring planting may prevent newly seeded clams from being preyed on by the resident

Plastic mesh netting being installed on a beach prior to seeding the beach with young clams. The netting often greatly improves the survival of such seed, but is most practical only for small plots not subject to intensive commercial harvest. (Photo courtesy K.K. Chew)

population of moon snails, shiner perch, and ducks. Again, plastic netting placed over the beds may substantially improve the yields from both naturally spawned and artificially seeded clam beds, particularly where predation is a significant problem.

**Intertidal Harvesting**

Most Puget Sound clams are harvested from the intertidal zone. Shallow burrowing clams (no deeper than about eight inches) such as native littlenecks, Manilas, and cockles are raked out of the ground with a spading fork or a long-tined or ordinary garden rake. The clam fork is a short-handled tool with eighteen-inch-long tines bent at a ninety-degree angle, which makes it easier for the digger to turn over the gravel-sand mixture where clams burrow. After the clams are dug, they are usually held in containers (either in "sink floats" over the beds or on land in trays supplied with flowing seawater to clean them of sand before they are sacked or boxed, or until there are enough to ship.

The amount of intertidal ground turned over by the average digger is about 20 to 40 square yards in the usual four-hour period on normal low tides. Generally, a minimum concentration of four to five pounds per square yard (or about 200 clams at the minimum commercial size of about 1½ inches in length) is necessary to provide modest wages to the diggers. In some productive Manila clam grounds, it is not uncommon to take as many as 500 marketable clams per square yard.

Horse and butter clams usually are found in the same habitat as native littlenecks, but because they burrow much deeper they are more difficult to harvest by hand. The same harvesting problem also applies to softshell clams, which are even more difficult to hand harvest because of their fragile shells. Because horse and butter clams are most abundant in lower intertidal and subtidal areas, the most efficient way to harvest them is with mechanical harvesters (see below).

An alternative tool, the hydraulic rake, has been proposed by one operator for softshell clam harvest in Port Susan and Skagit Bay. The hydraulic rake is used on intertidal gounds at low tides (when the clam beds are exposed). It consists of a hand-propelled unit with a manifold containing six or more water jet nozzles. Water is pumped to the manifold from a motor pump mounted in a small boat. These tools are labor saving, efficient, and appear to harvest clams with minimal breakage, but it is not known whether they alter the substrate or damage smaller clams, plants, and animals.

**Subtidal Harvesting**

Subtidal clams such as horse clams, butter clams, and geoducks are collected with mechanical harvesters or by divers using hand-held wa-

Hand dredging for Manila clams in Little Skookum Inlet. The tool here is a long-tined rake. Each bucket holds about 40 pounds of clams.

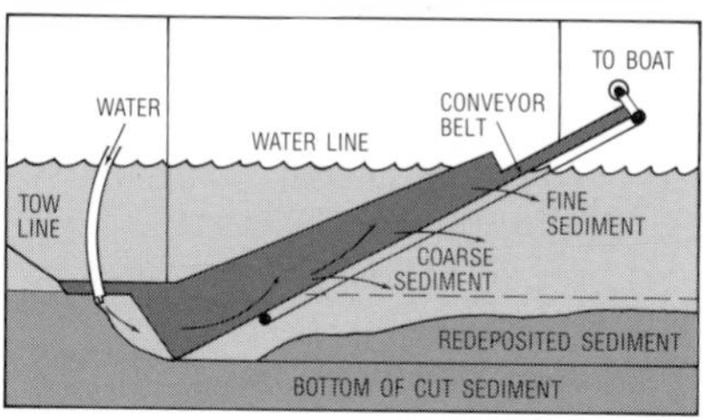


Tracks left in the mudflat at Port Susan, near Camano Island, shortly after the area was harvested with an hydraulic escalator (insert). The harvester loosens the clams and sediments with water jets, and the clams are then carried to the surface on a conveyor belt. (Photo courtesy Washington Department of Fisheries)

ter jets. The mechanical harvester that has been used in Puget Sound is a hydraulic escalator (also called "Hank's" machines after the inventor), which was developed on the East Coast to harvest softshell clams. The harvester shoots jets of water to loosen the substrate ahead of a cutter blade. The loosened gravel and sand is washed over the blade—clams and all—onto a conveyor belt, which carries gravel, shells, clams, and other materials to the surface. The clams are then picked off the belt, and the debris is allowed to fall back into the water.

Hydraulic harvesting of subtidal stocks in Puget Sound began in the early 1960s, but there are now only two hydraulic harvesters operating on leased hardshell clam beds in northern Puget Sound. In the late 1970s as many as ten machines were operating in the Sound, but poor market conditions, increasing government regulations, adverse public reaction to large, noisy equipment, and environmental constraints combined to drive the hydraulic harvesters out of the Sound. Hydraulic harvesters were also used to gather softshell clams, but have been restricted as a result of opposition by the Washington Department of Game, which claimed that dredging harmed intertidal flats and disturbed waterfowl, and by adjacent waterfront landowners, who objected to the noise, alteration of the intertidal land, and disruption of their views.

Another subtidally harvested species, the geoduck, is restricted to divers harvesting the clams one at a time using hand-operated water jets. The water jet loosens sand and gravel around the clams until they can be plucked from the bottom by hand. The jet leaves a small hole in the bottom, but immediately after the clam is removed, the hole begins to fill as the surrounding high-fluid-content sediment slumps into the void. There are no restrictions based on seasons or sizes for geoducks taken from lease tracts, but harvest is limited to daylight hours.

## Clam Growth and the Environment

The intertidal zone is part of a dynamic and changeable environment where winter storms, floods, and man's activities can alter substrate through erosion and deposition of silt. Clams that grow in these areas can be removed by erosional forces in the same way as beach materials, or can be smothered by new layers of silt. Even in deeper water where wave forces are weaker, subtidal clam beds can be affected by sediment motion associated with heavy storm waves or silt-laden runoff.

### Pollution and Red Tides

The main cause of clam bed closures to commercial harvest has been due to coliform bacteria contamination; except for PSP, other sources of contamination play little part in clam bed closures. The problem of shellfish contamination is not insignificant; for example, about 25 percent of available geoduck tracts cannot be leased because coliform counts are excessive or the tracts are too close to STP outfalls. These restrictions are enforced to protect human health and have no bearing on the effects of pollutants on the clams themselves.

Affected intertidal clam growing areas that do not meet certification standards of the Washington Department of Social and Health Services (DSHS) include Dyes Inlet and parts of Oakland Bay, and Port Susan (see Table 2.1).

The closure of clam grounds due to coliform contamination may be permanent or can occur only during and shortly after rainstorms, or result from breakdowns in sewage treatment plants. Sewage treatment plants and the encroachment of large-scale development have, historically, been major factors responsible for the closure of clam beds. A treatment plant that is constructed with a marine outfall results in the closing to commercial shellfish harvest of all waters inside a half-mile radius of the outfall. The closed area is a zone considered subject to sewage effluent presence that may expand at particularly large outfalls or change shape when currents move sewage alongshore (see Chapter Two for further details about coliform bacteria contamination).

Controlling sewage will not solve all pollution problems in clam growing areas. Water quality surveys indicate that stormwater runoff is an important contributing factor. Evidence points to stormwater as a principal source of coliform bacteria in urbanized areas, and to failing septic systems, dairy farms, and "hobby" farms in rural areas. A study of New York's Long Island Sound showed the predominant cause for withholding certification for about 25 percent of all shellfish beds was coliform bacteria deposited by urban stormwater runoff. Closures do not necessarily mean all clams in the affected beds are unfit to eat but it does prohibit commercial harvest for human consumption.

Like oysters, clams are most susceptible to pollutants, such as industrial chemicals, pesticides, and petroleum products, during embryonic and larval development. These materials may pose a human health hazard and some can accumulate in sufficient quantities to reduce the growth rates or even kill the animals or plants living on the seafloor. For example, on the East Coast mortalities have been attributed to clams burrowing in oil-soaked substrate lying below a seemingly clean surface sediment. The softshell clam industry of Puget Sound may be particularly sensitive to sewage-borne pollutants and industrial pollution because clam beds are usually near rivers and streams, which carry accumulated loads of industrial pollutants and agricultural runoff from upland areas.

Efforts to reduce shellfish mortalities and accumulations of toxic materials and bacteria are aimed at source controls and procedures to cleanse or "depurate" the animals. Reductions in waste loading and effluent discharge do reduce the toxicity of receiving water to oyster and clam larvae. Dramatic and significant improvements in shellfish survival occur when prescribed methods of wastewater treatment are used by pulp mills, oil refineries, and other industrial dischargers. Water quality problems associated with sewage outfalls can be alleviated by prohibiting effluent disposal in areas ideally suited for clam culture, or by forming sewer districts that minimize the number of sewer discharge points and improve the quality of the effluent.

It is also possible to remove accumulated toxins or bacteria by holding the shellfish in uncontaminated waters. This method is commonly applied to those animals grown in "classified" waters having elevated bacteria levels. The usual practice in Puget Sound is to "relayer" clams and oysters on uncontaminated beds until bacterial counts in the shellfish are reduced to an acceptable level. Holding the shellfish in tanks or ponds containing flowing uncontaminated or purified seawater (depuration) is also an acceptable procedure, but is an expensive process and has not yet been used by Puget Sound shellfish

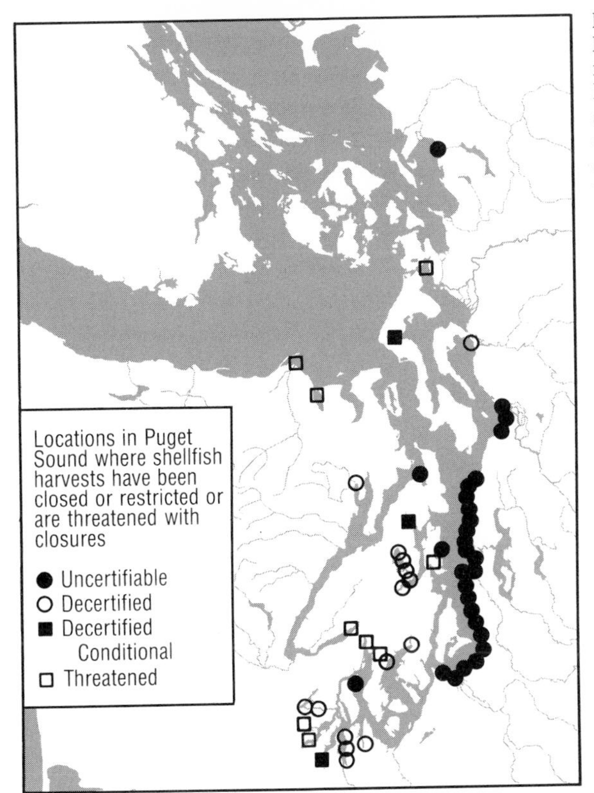


Locations of areas within Puget Sound where commercial shellfish harvest has been closed or restricted due to bacterial contamination. These areas are commonly associated with urban centers, marinas and sewage outfalls.

growers. In all cases, strict sanitary standards ensure that the depurated bivalves meet public health requirements.

Paralytic shellfish poisoning (PSP) has caused harvesting closures on the Strait of Juan de Fuca and as far south as Des Moines in Puget Sound (during the summers of 1978 and 1979). During a PSP outbreak, clams become contaminated and intertidal beds are temporarily closed. Hardshell and geoduck clams are not usually as seriously affected as mussels. Detoxification may take many months.

In short, nearly all pollutants threaten the clam industry in Puget Sound by increasing public health hazards and reducing the availability of the resource to the commercial harvester. Pollutants also threaten the future of the resource because some of them are detrimental to the survival of larval, juvenile, and even adult clams.

**Predators and Disease**

Because of the burrowing habits of clams, they are often protected from many of the predators found on oyster beds. This is generally true only for the adults; however, the larvae face many of the same predators as other marine larvae.

The moon snail, *Polinices lewisi*, is a major clam predator in Puget Sound, attacking clams burrowed in the substrate. Moon snails are generally found in the lower intertidal zone preying on butter clams and native littlenecks. Cockles and horse clams are rarely attacked because cockles have thick shells and horse clams burrow deeply. Manila clams are also not usually affected because of their location higher in the intertidal zone. In southern Puget Sound, moon snails are a significant problem, and the only effective way to control them is to remove them by hand. It should be noted that moon snails are edible and thus it may be possible to find a specialized market for them, which could make up for some of the expense of removing this unwelcome pest.

Other significant predators on smaller and shallow burrowing clams are crabs, fish (such as flatfish and rays), and diving ducks, particularly surf scoters. Plastic-mesh netting laid on the beach may help protect clams from many of these predators.

Clams in the natural environment are affected by few identifiable diseases, but as aquaculture and rearing techniques are improved, diseases common to high-density populations could become more of a problem. Bacterial and fungal diseases have occurred in hardshell clam hatcheries. Southern Puget Sound growing areas experienced heavy mortalities in a "winter kill," which may have been related to a combination of low salinities, freezing water temperatures, and microbial diseases.

## Resource Assessment and Production

Abundant U.S. clam resources account for about one-third of the total world clam harvest and total landings have increased from less than 40 million pounds during the 1940s to over 108 million pounds in

The moon snail, with a shell of up to 4½ inches in height, is a carnivore and feeds on clams by boring a large hole through the shell and sucking out the meat. It is common on sand flats in Puget Sound and produces a distinctive sand-covered and collar-like egg mass.

1982. About two percent of the total U.S. commercial clam production, and 18 percent of the hardshell clam production, results from private aquaculture. Washington State accounts for about 80 percent of the total U.S. West Coast hardshell clam production, and nearly 100 percent of the geoduck harvest. There are several reasons for the state's lead on the U.S. Pacific Coast:

- there is a long tradition of private ownership of tidelands in Washington, which has encouraged commercial harvesting;
- commercial production in Oregon and California is relatively low because these states place a much greater emphasis on recreational clamming, and available clam grounds are limited in those states;
- the Alaska clam industry is plagued by recurring red tides (PSP) and high labor costs.

Most of Washington's commercial clam harvest occurs in the bays of Puget Sound. There are at least 21 commercial clam farmers who own or lease about 680 acres of tidelands in Washington State. The larger farms harvest more than 150,000 pounds per year. There is only limited production of Manila clams in Willapa Bay, and the coastal razor clam catch is primarily recreational. Until the 1960s, the coastal razor clam fishery accounted for a large share of the commercial clam catch in Washington, but razor clam stocks and commercial harvests have since declined significantly.

Puget Sound harvests are generally adequate for regional markets, but increased harvests would be required to meet the market for clams taken elsewhere in the United States. Total hardshell production (excluding geoducks) in the Puget Sound region has remained relatively stable at one to two million pounds (total weight) per year, except dur-

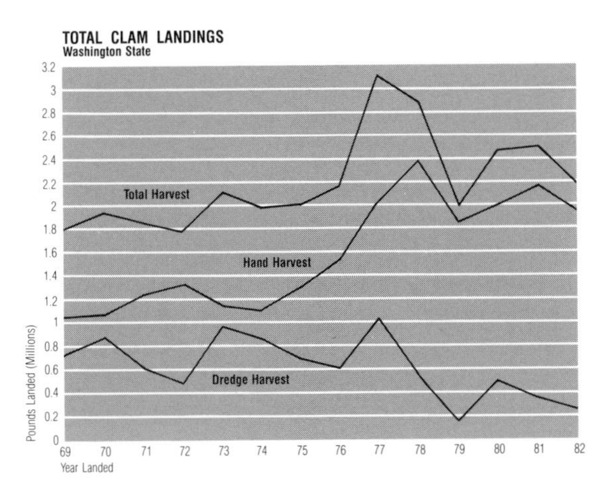


Total hardshell clam production (in whole weights) in the state of Washington (includes Puget Sound, Willapa Bay, and Grays Harbor)—Puget Sound landings accounted for 99 percent of the total. Geoduck harvests are not included—they were about five million pounds in 1982. (Source: Annual landings data for 1969–82, Washington Department of Fisheries)

ing the late 1930s, when about three and a half million pounds were harvested.

Subtidal lands fall under the public domain, therefore subtidal hardshell clam harvesting is managed and controlled by the state. As of 1982, only about ten percent of the hardshell clam harvest was from state-controlled submerged lands. Following are production summaries of the major individual clam fisheries—both intertidal and subtidal—by species.

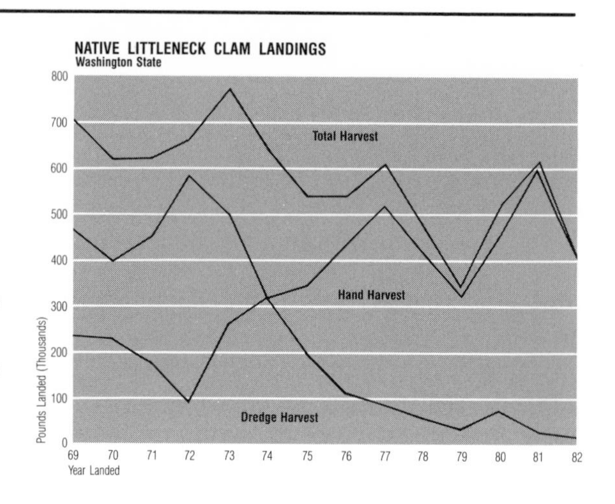


Puget Sound native littleneck clam landings and aquaculture production (in whole weights). Dredge harvest in 1982 was less than 5,000 pounds. (Source: Annual landings data for 1969–82, Washington Department of Fisheries)

**Native Littleneck Clams**

The production of native littleneck clams declined in the 1960s and 1970s from approximately one million pounds per year in the late 1950s and early 1960s to 500,000–700,000 pounds in the late 1970s and early 1980s. In the early 1970s a large proportion of the landings were from mechanically harvested subtidal stocks. Recently, these harvests have declined to less than ten percent of the intertidal hand harvest. Subtidal native littleneck stocks are estimated at up to 28 million pounds for Puget Sound waters.

**Manila Clams**

The Manila clam fishery began making significant contributions to the hardshell clam fishery after World War II. Production has ranged from about 300,000 pounds per year in the early 1960s to over 1.5 million pounds in recent years. Virtually all Manila clam production occurs on intertidal beds in Oyster Bay, Totten Inlet, Eld Inlet, and Little Skookum Inlet. With the exception of the geoduck, the Manila clam is the most valuable commercial clam resource in Washington.

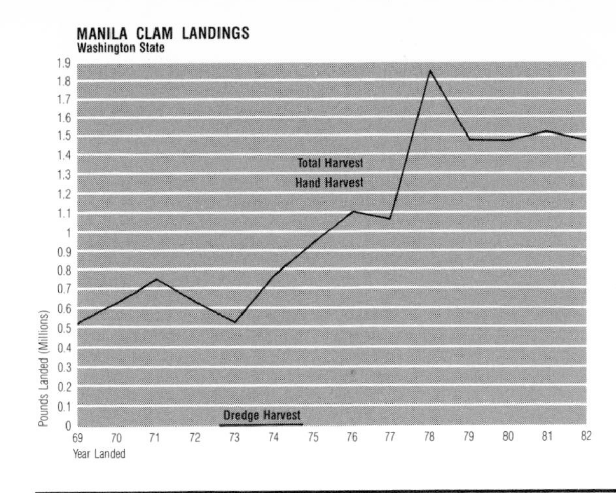

Puget Sound Manila clam production (whole weights)—mostly aquaculture production. Dredge harvest in 1982 was about 1,500 pounds. (Source: Annual landings data for 1969–82, Washington Department of Fisheries)

## Butter Clams

The hand harvest of butter clams has declined since 1960 because of the entry of mechanical harvesting, relatively low prices paid for the clams, and the fewer clam beds suitable for commercial production. Since 1969 up to 570,000 pounds per year were harvested from subtidal beds, and production costs were less than for intertidal hand-digging. Subtidal stocks are believed to be substantial, with an estimated standing crop in Puget Sound of 114 million pounds.

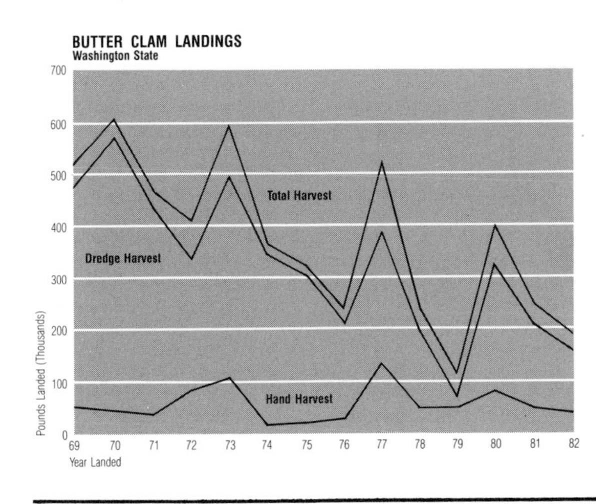

Puget Sound butter clam harvest (whole weights). Dredge harvests dominate the recent landings. (Source: Annual landings data for 1969–82, Washington Department of Fisheries)

## Horse Clams

There has never been an extensive commercial hand harvest of horse clams in Puget Sound, and although small numbers are taken by

Puget Sound horse clam landings (whole weights). Present production is limited and almost all clams are now dredge harvested. (Source: Annual landings data for 1969–82, Washington Department of Fisheries)

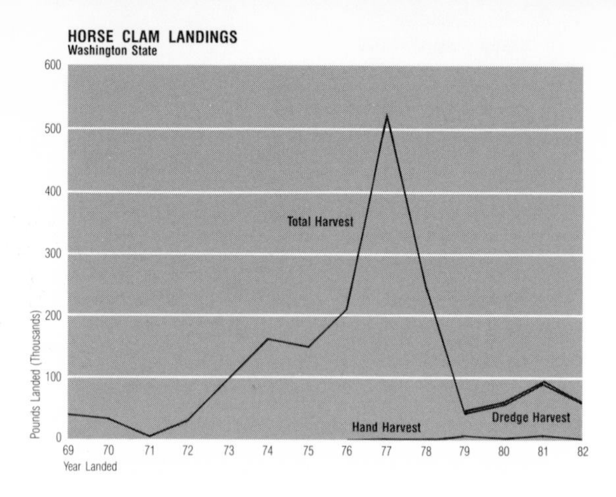


mechanical harvesters, the price has been, until recently, too low to sustain a profitable operation. Mechanical harvester landings of horse clams exceeded 520,000 pounds in 1977; however, since 1979 landings have been less than 100,000 pounds per year. Horse clams are frequently mixed in with butter and native littleneck clams, and the standing crop estimate of these large clams in Puget Sound is 29 million pounds.

---

Puget Sound softshell clam harvest (whole weights). Present production is restricted to hand harvest—there is no dredge harvesting. (Source: Annual landings data for 1969–82, Washington Department of Fisheries)

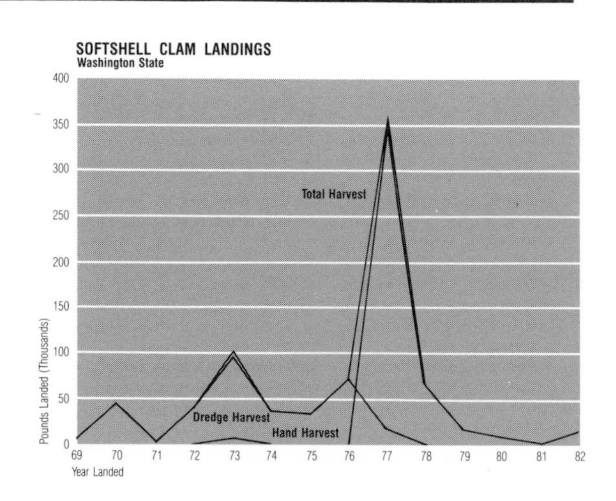


## Softshell Clams

The softshell clam industry has the potential of being a significant Washington State clam fishery. Surveys of commercially harvestable areas in Port Susan and Skagit Bay indicate a standing crop of about 18.7 million pounds covering 1,200 acres, with an estimated annual maximum sustained yield of about 1.9 million pounds. Commercial harvest of softshell clams using a mechanical harvester began in 1969 on privately owned and leased intertidal grounds. These operations ceased in 1977 amidst a flurry of lawsuits and citizen complaints. The hand harvest of softshell clams has continued, and in 1977 over 340,000 pounds were harvested. Only 12,200 pounds of softshell clams were harvested in Puget Sound in 1982.

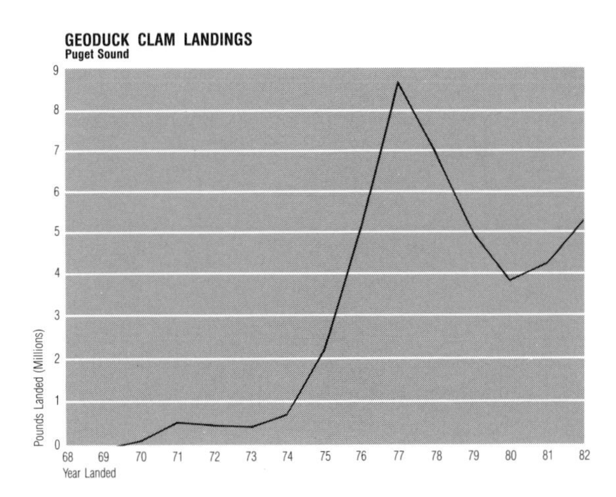


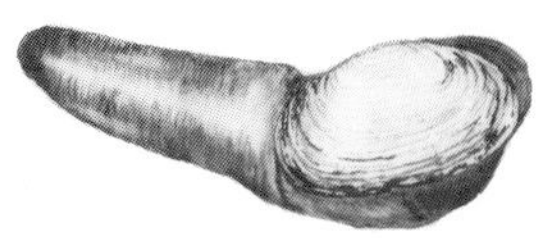

Puget Sound geoduck clam harvest (whole weights). Present production is about five million pounds. There are no records of commercial harvest prior to 1970. (Source: Annual landings data for 1969–82, Washington Department of Fisheries)

## Geoducks

The geoduck is plentiful in subtidal areas. The Department of Fisheries estimates there is a standing crop of 117.6 million geoducks in the 33,799 acres surveyed through 1977. Major beds totalling 15,507 acres were estimated to contain nearly 100 million geoducks. These estimates are thought to be conservative because of incomplete surveys in all areas of the Sound.

There are now over ten geoduck tracts totalling about 1,200 acres leased by the state for harvest in Puget Sound. Typical locations are like those in Drayton Passage and Useless Bay, with one to two boats and four to six divers per boat operating on each tract. The average harvest is about 6,000 pounds per acre per year. The harvest per diver-day ranges as high as 400 to 500 pounds (whole wet weight).

The geoduck industry consists of a small number of private operators committed to harvesting, processing, and marketing their product. The meats are sold primarily outside the state: neck meat goes to overseas markets in Japan and Taiwan or to U.S. markets, and body meat is sold in California and on the East Coast. There is a market for clam trimmings, but the supply often exceeds the demand.

In 1970, 81,000 pounds of subtidal geoducks were harvested. Catches increased markedly in the years that followed, with production in 1977 of 8.6 million pounds. One large processor reported landings averaging 20,000–25,000 pounds daily from March through October of 1977. The industry is now limited to a harvest quota of about five million pounds per year.

## Clam Economics

### Costs

The economics of the clam industry are complicated because of its diversity and wide variation in production costs, efficiency of harvest, and earnings per pound. For example, in 1983 the costs (per 1,000 pounds) of harvesting and marketing Manila clams on an intensively managed South Sound farm were as follows:

Harvest labor $340
Handling, shipping, and packing $23
Transportation $23
Insurance $13
Maintenance $6

In addition, fixed costs, such as farm management, grounds surveys, seed, etc. add up to about $3,400 per month for a farm production of 100,000 pounds per year. A clam farmer who leases private or state land also pays a stumpage fee based on pounds harvested, which ranges from three to five cents per pound for state lands to ten to twenty-five cents per pound for private land. Combined royalty fee and bonus bid prices for geoduck tracts range from nine to nineteen cents per pound.

### Income

The highest prices are paid for the steamer clams. Prices in 1981 averaged $1.01 per pound wholesale and ranged from $.68 to $1.25 per pound depending on sales volume and sales channel. Growers who sell directly to the restaurant/retail sector generally receive higher prices (e.g., $1.20 to $1.25 per pound in 1981) than growers who sell clams to wholesalers. As of March 1985, domestically produced Manila clams were averaging $1.10 per pound on the open market. Clams which require processing, such as the butter clam, bring in very little to the producer.

### Markets

Puget Sound clams are marketed live or frozen in the shell for use as steamers, or the shucked meats are marketed fresh or canned. There is a steady demand for steamer clams in the shell, and available supplies are readily sold at high prices. Native littleneck clams, Manila clams, softshell clams, and small butter clams are considered ideal steamers. Manila clams are often preferred over native littleneck clams and they have a longer shelf life. Large butter clams, cockles, and horse clams, which tend to be somewhat tough, are usually shucked and the

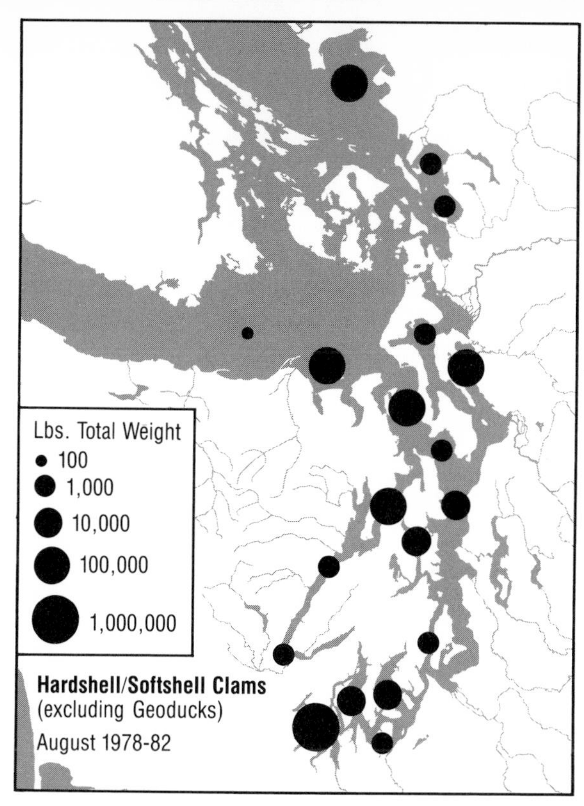


Distribution of hardshell and softshell harvests and aquaculture production in Puget Sound, 1978–82. Each circle depicts average annual landings from the value given up to the next highest value.

meats are sold fresh or canned for chowder. Softshell clams may also be shucked and sold either fresh or canned for institutional use or at the retail level. Other markets for hardshell and softshell clams include a small (but expanding) half-shell trade, clam strips, and fish and crab bait.

Canned clams from the Pacific Coast must compete with canned surf and softshell clams from the Atlantic Coast and with canned clams imported from Japan and Korea. Because the East Coast and imported clams are low priced, canning is only a marginally economical venture on the Pacific Coast. Large populations of surf clams discovered a few years ago off the Alaska coast may eventually prove to be competition for Washington suppliers and may augment declining stocks of East Coast surf clams.

The local market for fresh hardshell clams is also influenced by imports. Quantities are harvested from natural populations in British Columbia, and there is significant potential for harvest in Alaska. On a seasonal basis substantial quantities of fresh steamer clams are imported from British Columbia at prices as low as $.65 per pound. Large numbers of Alaska hardshell clams may also become available if a reliable and safe procedure is developed to detoxify clams exposed to PSP,

Distribution of geoduck harvests in Puget Sound, 1978–82. Each circle depicts average annual landings from the value given up to the next highest value.

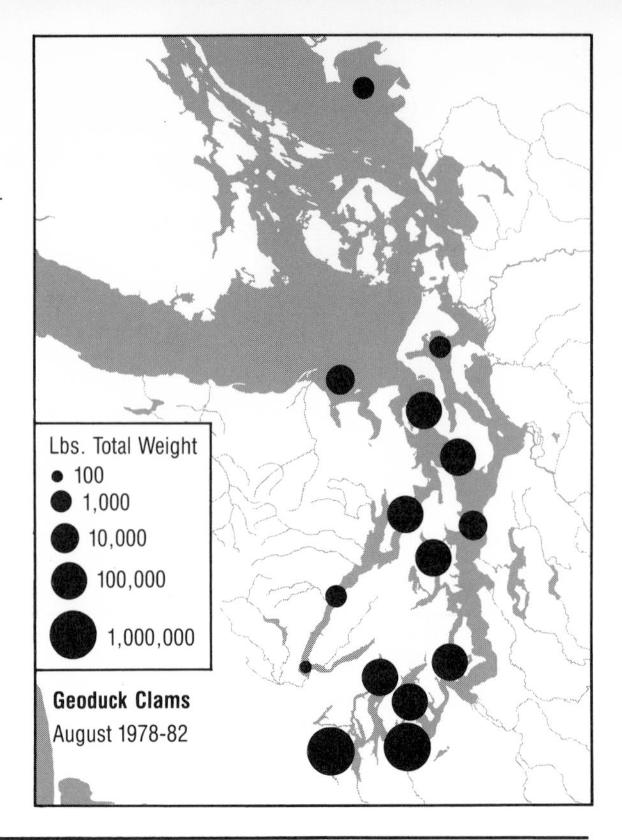


a common problem in Alaskan clams. No fresh clams are imported from countries other than Canada.

The most recent innovation in the Puget Sound clam industry was the rapid development of the commercial geoduck harvest in the late 1970s. The product is sometimes marketed as "king clam steaks" (to overcome consumer reluctance to buy geoduck). The tougher portions are ground, shaped into patties and frozen, or canned for use in chowder. The product has gained market acceptance, as processing techniques have improved and demand increased, despite high harvesting and processing costs.

## The Future of the Clam Industry

The culture and harvest of clams in Puget Sound has great expansion potential provided certain conditions can be met. The limited availability of habitat suitable for the culture and harvest of clams is a prime constraint to expanded production. Because the state sold or leased tidelands for oyster growers and to owners of adjoining uplands, most of the tidelands within the bays and estuaries of Puget Sound are already privately owned or leased. Widespread commercial clam digging on public land is not possible within Puget Sound, because there are few productive public areas available within the bays, and most of them have been set aside for a growing recreational clam fishery. The land issue can be offset to a great degree by more intensive management and enhancement of existing commercial beds. This requires the application of good farming practices—stock assessment, soil improvement and stabilization, and artificial seeding.

Water quality and the specter of decertification will continue to be a serious concern. While it is hoped that a constant improvement in pollution control will lead toward improving water quality, it has become increasingly obvious that shellfish growers and harvesters will have to take a more active role in encouraging proper water quality management (see Chapter Two for further details).

Poaching on privately owned or leased land is a recurring headache for intertidal clam farmers and the only way to prevent it is constant patrol and surveillance of clam beds. In some cases restriction of clam bed access helps, but poachers can easily approach by boat. The posting of private property to distinguish it from adjacent public land does little to deter trespassing by recreational clam diggers. Attempts by the state to increase access to public beach lands only complicates the matter unless the public is willing to honor private ownership of the resources and restrict their activities to publicly owned lands.

**Technological Advancements**

Clam harvesting is a labor-intensive industry. Both geoduck and intertidal clams are hand harvested. Hand harvesting is efficient and profitable in areas with relatively high clam densities. Although some mechanical harvesting has been done in intertidal beds for softshell clams, much of the hardshell clam ground is divided into areas too small for mechanical harvesting or too rough for digging other than by hand. It is therefore unlikely that the traditional harvesting methods for either intertidal clams or geoducks will be greatly altered in the near future.

Development of the hydraulic escalator harvester provided an effective means to exploit previously inaccessible subtidal and marginally profitable, low density clam beds. But the hydraulic harvester is adapted to work effectively only in relatively shallow water (less than 25 feet) and on specific types of substrate. It is also thought to cause long-term alteration to subtidal clam beds. For these reasons and because of the lack of acceptance by upland owners, development of Puget Sound's deep water clam resources appears unlikely. Although the Washington Department of Fisheries has located some deeper water clam beds (generally, below 100 feet) having a reasonable population of harvestable clams, most of the population consists of very small clams. However, improved processing technology—combined with development of additional markets—could increase the use of these small clams and clam parts presently underutilized or discarded as waste.

**Regulation and Use Conflicts**

Aquaculture ventures must compete with many other public and private uses of limited space on a rapidly shrinking shore. Waterfront residents have aesthetic objections to hydraulic harvesters, barges, poles that mark the corners of beds, and other paraphernalia associated with clam farming. Clam farmers using intertidal beds have also expressed their concern about potential adverse effects of mechanical harvesters on clam ground productivity. They believe that hand-digging tends to improve clam beach quality and that mechanical harvesters tend to cause substrate compaction.

Intertidal clam farming is not highly regulated, although it is taxed at a higher rate than oyster farming. Also, it may—like oyster culture—require modification and maintenance of beds to remain competitive. Mechanical harvesting of clams and geoducks, on the other hand, is intensely regulated. The regulatory climate is especially frustrating because it is subject to varying interpretations. Moreover, procedures required for entering the subtidal clam and geoduck fishery are complicated and lengthy, and the outcome is uncertain.

Prospective clam harvesters interested in opening up new areas for shellfish culture or harvest can expect to take up to twelve to eighteen months or longer to complete the permit and review process. The steps involved in this process are as follows:

1. Obtain a lease for the harvest tract from the Department of Natural Resources if the tract is subtidal or is located in public tidelands.
2. If a mechanical harvester is to be used, obtain a harvesting license from the Department of Fisheries.
3. Comply with provisions of the state Shoreline Management Act as administered by county agencies.
4. Obtain a combined Section 404 and Section 10 permit from the U.S. Army Corps of Engineers.
5. Obtain a sanitation certificate of approval from the Board of Health, Department of Social and Health Services.

At each step, the environmental considerations of harvesting are weighed, and in many cases the added burden of an impact statement or assessment may be required.

Subtidal clam harvesters who pass the preliminary hurdles and begin operations face yet another major obstacle to success. Upland residents are challenging the smaller, less cohesive industry group. Confrontations have occurred at Agate Pass between Bainbridge Island and the Kitsap Peninsula; at Port Susan, inside Camano Island north of Everett; Wescott Bay on San Juan Island; and Penn Cove on Whidbey Island. The conflicts in Agate Pass have centered around degradation of the aesthetic environment by stirred-up sediment plume formation and the obstruction of view; machinery noise; trespassing of the dredge onto intertidal and private clam beds; wastage; and impact on important incidental species, and other biological and management-related problems. In addition, sports fishing groups are concerned that the clam harvests may adversely affect salmon migration and bottomfish populations. These conflicts have shut down harvesting in Agate Pass and, although the Shorelines Hearing Board attempted to resolve the key issues and found in favor of the harvesting, the matter was subsequently taken to the Superior Court by the county.

Delaying actions have had a substantial economic impact on the clam harvester in Agate Pass and undoubtedly discourage other operators seeking to enter this field elsewhere in Puget Sound. The Shorelines Hearing Board recommended in the review of the Agate Pass case that the Department of Fisheries conduct closely monitored tests of mechanical harvests in both subtidal and intertidal beds. There are no available long-term data on recovery rates of mechanically harvested clam beds, the actual yields that can be sustained without depleting the beds, and the interaction of subtidal clams with intertidal stocks.

**Market Development**

Of the Puget Sound clams that reach the domestic market, there are two source groups noted above; subtidally harvested clams and intertidally cultivated clams. The advantages of intertidal cultivation will continue due to consistent availability and quality. Subtidal harvesters will remain dependent on wild seed production at least until geoduck seeding becomes operational. Therefore, the vagaries of natural repopulation will determine the amount of product available to them. Intertidal cultivation will continue to maintain a high degree of control over the standing population of clams available for harvesting, by artificial seeding and proper management. The size of the harvested clam can also be regulated by hand harvesting. Locally produced clams should be able to compete favorably with foreign imports by offering a better quality product with longer shelf life. East Coast clams cannot compete with local clams in the fresh market. This advantage applies particularly to the West Coast and Hawaiian markets. Canadian clams (i.e., Manilas from British Columbia) will continue to exert a strong seasonal influence on prices and will be a major competitor.

Clams will continue to be an attractive commercial resource because there is a consistent market and prices are increasing. New farming techniques and the development of additional hatcheries will make it possible for clam production to expand, provided that suitable growing and harvest areas are available, water quality is not degraded, and the necessary permits can be obtained.

*The Octopus Catcher* by Edward S. Curtis, c. 1914. A Kwakiutl man of British Columbia pulling an octopus to shore. Octopus and other shellfish were collected by hand or nets and formed part of the diet for coastal Indians throughout the Pacific Northwest. (Photo courtesy Special Collections Division, University of Washington Libraries)

CHAPTER FOUR

# Mussels and Other Molluscs

There are several species of marine molluscs besides clams and oysters that play a role in Puget Sound's shellfish industry. Octopuses are used for food and as bait. Increasing numbers of squid are taken and are beginning to be found on the menus of seafood restaurants and in retail outlets on the Pacific Coast. Two common species of mussels, the blue or bay mussel, *Mytilus edulis*, and the California mussel, *M. californianus*, are found in Puget Sound. The blue mussel is the major species of commercial interest in Puget Sound, although since 1980, the California mussel has been harvested from the outer coast of Oregon and introduced to markets across the country.

Two other mussel species, *Modiolus capax* and *M. rectas*, commonly called horse mussels, occur here but are not abundant and have no potential as yet as commercial species. In addition, several species of scallops and abalone are harvested commercially or are being evaluated for commercial aquaculture.

## MUSSELS

Considered an important seafood commodity in Europe, mussels have been cultivated in France for over 700 years. Mussel farmers in Spain, the Netherlands, and other coastal countries in Europe produced over 400,000 tons in 1981. Intensive culture of mussels in those countries produces the greatest yield of protein, per unit area, of any type of animal husbandry known.

In North America, mussels were common in the diets of Indians and white explorers, but until recent times, have not been popular or extensively harvested. Interest in mussels as a source of protein increased with the onset of World War II because mussels were not rationed and were freely available. From 1942 to 1947 up to 1,350 tons were harvested annually in the United States, but after the war consumers returned to more traditional seafoods, which—combined with fewer available high quality mussels—reduced production to a fraction of those levels.

Resurging interest in mussels in the United States since the mid-1960s has led to increased harvesting and cultivation, with production reaching about 7,500 tons in 1981. Much of this production now takes

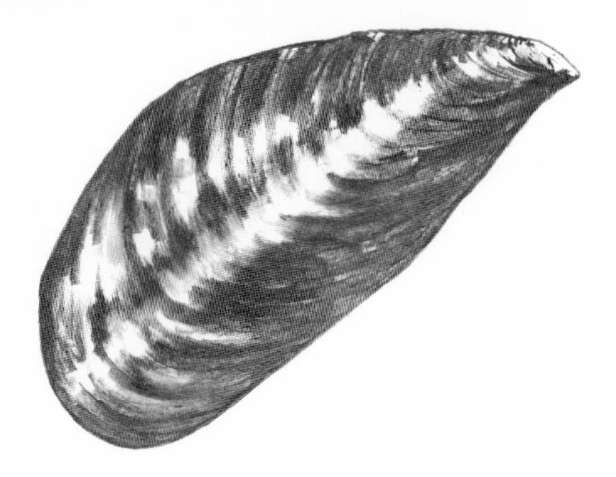

The blue mussel is easily recognized by its smooth shell, hinged at the end. It attaches itself by hairlike threads called a byssus to just about any stable surface from subtidal to high intertidal zones. Frequently, the shells become encrusted with barnacles and algae.

place on the East Coast, particularly in New England. Several growers in Puget Sound are now beginning to make a small contribution to the total U.S. mussel production, after a shaky start in the 1960s when commercial endeavors were limited by uncertain markets and supplies.

## Seed Collection

Mussels have a life history that is similar to oysters or clams, and they spawn naturally in Puget Sound, usually in the spring and early summer. The developing mussel larva remains free-swimming for up to one month and then seeks out an object (usually covered with fine filaments) on which to settle. Settlement occurs throughout Puget Sound, but varies in intensity and location from year to year. The best settlement occurs near or below the low-water mark where the waters are somewhat confined and near an existing population of adult mussels.

All spat or seed used in the cultivation of mussels in Puget Sound are obtained from natural production. There is at present no artificial rearing—although seed production in hatcheries would be of significant benefit to the industry. Mussel farmers in Puget Sound frequently must contend with a shortage of seed arising from the unpredictable occurrence and ephemeral nature of spatfalls.

Spat for cultivation are collected on a variety of materials including synthetic rope and fibrous plastic netting. Spat may be collected by placing the collectors in racks held in the intertidal zones or suspended near the surface from rafts.

## Growing and Harvesting

Intensive mussel culture has only recently been initiated in the Puget Sound region using techniques adapted from European countries. It

Above: Intertidal pole-and-longline mussel farm located at Race Lagoon, on Whidbey Island. This system allows the farm to grow mussels on grounds that would be otherwise unsuited for culture. Left: Detail of an individual mussel-covered longline.

is possible to harvest mussels from natural intertidal stocks, which requires little capital investment, but natural stocks can have poor meat yields, discolored meat, and contain foreign substances (mud, for example) in the mantle cavity. Cultivation, however, produces crops of consistent size and flesh quality. The ability of mussels to attach and re-attach to various natural and manmade surfaces with strong elastic and adhesive threads (the "beard" or byssus) makes them very attractive for cultivation.

Because mussels are adaptable and hardy they are easy to cultivate using any of the following methods:

- bottom culture—as transplanted mussel beds;
- off-the-bottom culture—suspended from rafts, floats, or racks, on ropes or in tubes of plastic netting;
- *Bouchot* culture—attached to vertical posts, either by direct settlement or by binding seed on netting.

In any form of mussel cultivation, the aim is to grow crops of optimal size and meat content for harvest. To maintain steady production, three prime requirements must be satisfied: (1) a regular source of spat or seed stock; (2) a suitable area for fast growth; and (3) good husbandry, e.g. thinning-out and protection from predators.

**Bottom Cultivation**

Bottom cultivation has been practiced on a small scale in Puget Sound. In Europe, it is a major form of cultivation, particularly in the Netherlands. It is basically a very simple method. Small mussels are transplanted from natural beds to locations having favorable water currents and rich food supplies. Larger mussels may also be relayed to improve growth and meat content. Mussels raised on the bottom may reach market size within a year, with faster growth rates occurring in the low intertidal to shallow subtidal zones.

**Post (Bouchot) Cultivation**

This technique is practiced in its traditional fashion along the Biscay and Brittany coasts of France. The mussels are grown on numerous rows (bouchots) of six-foot high wooden posts placed in the low intertidal zone of bays and estuaries. The mussels either settle directly onto the posts or spatted ropes are wound around them. The method is extremely labor intensive and annual productivity is low, about 50 pounds per post. One British observer noted that bouchot cultivation is economically viable in France because the French attribute a unique flavor to these mussels, so they command higher prices.

**Off-the-bottom Cultivation**

Mussels can be grown attached to cord and ropes, or in net tubes which are suspended from rafts and racks or on longline systems supported with buoys or poles. Raft cultivation is very large-scale business in Spain, and in Puget Sound several growers now employ both raft and longline systems.

With rope culture, mussel seed is collected on synthetic ropes from wild stocks during spawning periods or is transferred from natural substrates. The ropes are up to 30 feet long and are wrapped with fine mesh and suspended from a raft. The ideal raft size or density of suspended stocks has yet to be determined for Puget Sound. (In Europe, rafts are up to 60 feet square and can support up to a thousand ropes.)

Mussels grow much faster and are more productive when cultured off the bottom than on tidal flats or seabeds. They can grow from half-inch seed to over two inches in only five months during the summer or in seven months from fall through spring. In Puget Sound, ropes yield harvestable mussels at densities of about seven pounds per foot, equivalent to Spanish production levels. Growth rates are greatest near the surface of the water and decline significantly at depths below six feet.

Yields of 60,000 to 200,000 pounds of mussels per raft per year are obtained from culture operations in Spain. Assuming an average meat yield of 25 percent of total weight, median production is about 410,000 pounds of meat per acre per year! A mussel of marketable size (about two inches long) should weigh about a third of an ounce in the shell and contain about a tenth of an ounce of meat.

Off-the-bottom cultivation in Puget Sound does have several disadvantages. Moorings are expensive and may be difficult to emplace at suitable locations. Where upland landowners pose aesthetic objections which may limit raft culture, a longline system of hanging mussel culture may be used. Longlines supported by low visibility plastic buoys support vertically hanging ropes for mussel settlement. A typical system, equipped with 440 twelve-foot ropes, may have an estimated capacity of 45 tons per longline.

The principal advantages of off-the-bottom cultivation over bottom cultivation are avoidance of crab and starfish predation (although birds pose a serious predation problem) and rapid growth. The method is flexible and can easily be altered to varying depths and seabed topography. Adequate shelter and water flow are required, but areas are se-

Hanging culture is a solution to growing shellfish in a limited area. Suspended from these floats are lines holding strings of mussel-covered rope. This method is similar to raft culture, but is less costly and usually easier to maintain in open waters.

Above: A drum cleaner and declumper removes some of the animal and plant material that settles on growing mussels. Considerable hand labor is still needed, however, to remove the byssus, and to clean and sort the shells (right) to ensure a high quality final product.

lected for low turbulence and current speeds. In the best locations, spat settlements occur at or close to the cultivation sites; otherwise, collector ropes must be transported to cultivation sites in other areas. Net tubes also can be stocked with clumps of seed mussels stripped from spatted ropes.

Cultured mussels are not harvested until they are more than two inches long. Harvesting begins in late September and proceeds until late June, except just before, during, or immediately after spawning. Harvesting and processing are hand operations that are difficult and expensive to mechanize. The mussels must be shaken or picked from the ropes, size graded, and cleaned of algae and other fouling organisms that attach to the rope and shell.

## Environmental Considerations

A number of physical and biological factors are important to commercial mussel culture: high wave activity, salinity, suspended silt, and temperature extremes may contribute to mortality. The blue mussel, for example, is less firmly attached to the substrate than the California mussel and is more susceptible to being detached by waves.

Natural predators of mussels are sea perch, crabs, starfish, and birds; and competitors for food or space are barnacles, sea squirts, sea cucumbers, an occasional algae, and even an over-abundant set of seed mussels. Barnacles are a problem particularly when they set on ready-to-harvest mussels. Raft-cultured mussels raised close to the surface in Puget Sound are literally gobbled up by diving ducks, which can account for losses of up to 80 percent of potential production.

While the mussel farmer can usually combat predation and competition by marine organisms, it is very difficult to control predation by birds. Netting placed around rafts or floats is expensive, hard to maintain, and often ineffective. Loud noises (i.e., propane cannons) may work initially, but soon become routinely ignored. Shooting the birds is effective, but not necessarily uniformly approved. One grower con-

Starfish like this hungry specimen can be troublesome in the ground culture of mussels, but are generally not a major pest in hanging culture. Other predators, such as diving ducks, are much more destructive.

stantly patrols his farm during the day to drive off birds—a simple but time-consuming measure.

Diseases have not been a major problem in Puget Sound, but coliform bacteria contamination and PSP can cause decertification of cultured stocks. Mussels are generally more tolerant of environmental contaminants than other bivalves, but because they are filter feeders like oysters and clams, they ingest bacterial and viral pathogens, metals, and various hydrocarbons.

Similarly, the PSP toxin accumulates in the mussels, rendering them unfit for human consumption. Once the toxic phytoplankton (*Gonyaulax*) in the seawater die or drift away, the affected shellfish slowly lose their toxicity. Detoxification takes a few weeks to several months for mussels, and toxicity is monitored continuously by state inspectors and growers. So far, PSP has only affected mussels in north and central Puget Sound, and has not been a serious problem in the South Sound.

## Production and Marketing

Production of mussels by about five commercial operators in Puget Sound reached 95,000 pounds in 1981 with a wholesale market value of $110,000. The average price was $1.16 per pound, although mussels were sold directly to restaurants for up to $1.35 per pound. Sales increased somewhat in 1982 to 102,000 pounds. Puget Sound growers predicted sales in excess of 680,000 pounds in 1985, provided the necessary permits to increase facility size could be obtained. By contrast, mussel production in Europe is enormous: Spain alone produced over 100,000 tons (200,000 metric tons) from 4,400 rafts in 1981. Given fully realized raft potential equivalent to Spanish production, a single mussel raft

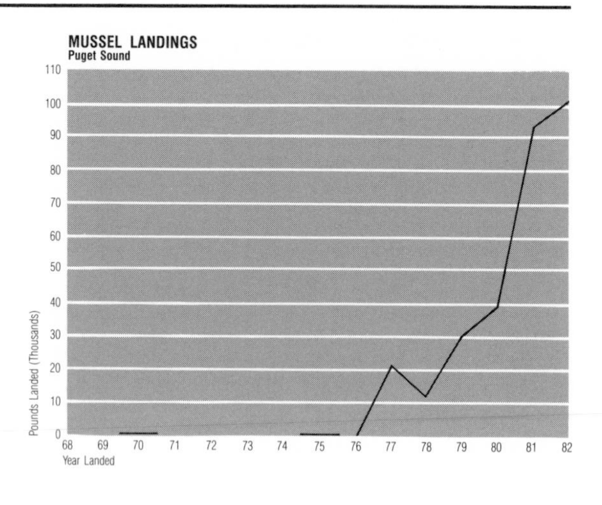


Puget Sound mussel landings (whole weights). Present annual production is estimated to be over 600,000 pounds. (Source: Annual landings data for 1968–82 reported by the Washington Department of Fisheries)

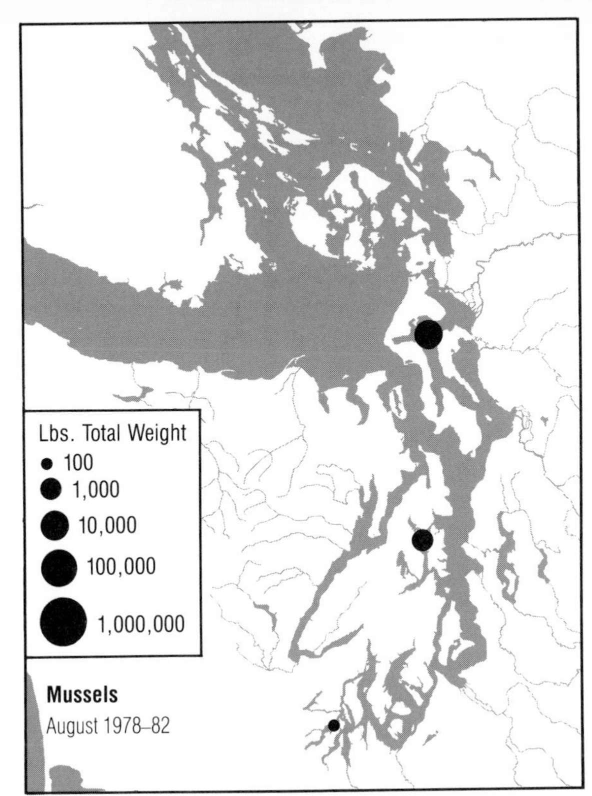


Distribution of mussel harvests and aquaculture production in Puget Sound, 1978–82. Each circle depicts the average annual landings from the value given to the next highest value.

operating in Puget Sound should produce a minimum of 60,000 pounds of mussels per year, representing a value of nearly $70,000 (assuming a price of $1.16 per pound).

All the mussels grown and marketed in Puget Sound are sold live and unshucked. Almost all are sold in-state due to limited production, with more than half being sold directly to restaurants. The demand for mussels grown in the Puget Sound area is substantial, but resistance to increased prices has led to some substitution of less expensive East Coast blue mussels for the Washington-grown product. Furthermore, the supply of good quality mussels has been limited and unsteady as the industry has developed. This has discouraged volume buying and dependence on a single source. With careful promotion, reliable supplies, and quality assurance, the market potential of mussels in the United States should improve.

Mussel culture technology can consistently produce a high quality product but production costs are high and the resulting product is expensive. Increased volume, consistent supply, and lower production and transportation costs would help Puget Sound growers compete more successfully with East Coast mussels. Stock losses and costs due to required reseeding should decline when effective predator control methods are developed. Development of improved mechanical mussel cleaning devices would also lead to lower labor costs.

New species and varieties can also be investigated. For example, experimental evidence suggests that the California mussel may be a potential culture species in areas of Puget Sound where the blue mussel often suffers high mortalities. Artificial rearing may be required to ensure a continuous, reliable supply of mussel seed. In addition, mussel farmers must work with growers of other shellfish crops and water resource groups to curtail water pollution. The use of "depuration" systems could be investigated as an alternative to decertification of prime mussel growing ground. Finally, the mussel growers must improve relations with shoreline residents and local planning commissions in order to expand operations in prime mussel growing areas.

## OCTOPUS

The octopus is the only other mollusc besides clams, oysters, and mussels that has had a history of a significant commercial fishery in Puget Sound. In the past, it was taken primarily as bait for the halibut fishery, but within the last ten years octopus has become more widely accepted as seafood.

Octopuses are found from the intertidal zone to deep subtidal areas of the Sound, where they inhabit holes and crevices. They are carnivorous, feeding on crustaceans, other molluscs, and fishes. It is estimated

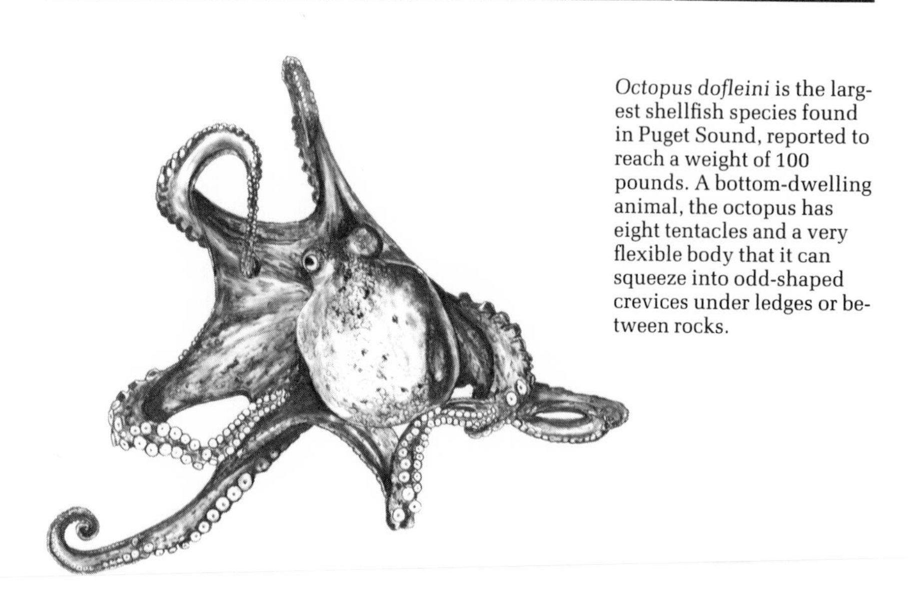

*Octopus dofleini* is the largest shellfish species found in Puget Sound, reported to reach a weight of 100 pounds. A bottom-dwelling animal, the octopus has eight tentacles and a very flexible body that it can squeeze into odd-shaped crevices under ledges or between rocks.

that there are perhaps 100 species in the octopus family in the world's oceans, two of which are common to the waters of Puget Sound, *Octopus dofleini* and *Octopus rubescens*. *Octopus dofleini*, the only commercially harvested species, can grow to be a giant, with an arm spread of up to ten feet and weight reaching 100 pounds. By comparison, *Octopus rubescens* rarely reaches a body length (exclusive of arms) of 2½ inches.

## Harvesting

A personal-use fishery for octopus has existed for many years near major population centers in Puget Sound. In the past octopuses were taken by hand or by gaff from the subtidal zone for markets in Japan or local Oriental communities, or to be used as halibut bait. A cloth bag containing copper sulfate crystals was tied to a stick and then thrust into a cave suspected of harboring one of the creatures. This forced the octopus into the open where it was gaffed. Both the gaff and chemical irritants are now illegal in Washington and cannot be used for any form of harvest, commercial or sport. Octopuses also may not be taken commercially by scuba divers, or by the use of such fishing devices as grapples, jigs, and lures, which are likely to mutilate the animals.

It is legal to harvest octopus with pots or traps, dipnet, or by hand, and a successful pot fishery existed near Dungeness Spit from the late 1950s until 1973. Octopus is also fished with pots near Tacoma and in Hood Canal. Presently, octopus is taken incidentally in traps used for finfish or in otter trawls.

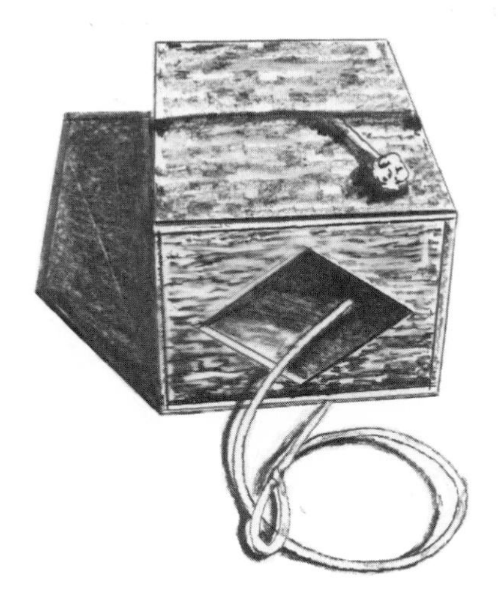

An octopus trap. This open-mouth type of box trap is usually made of cedar or plywood, 1 x 1 x 3 feet. Octopus traps are also made from plastic pipe, oil drums, plastic buckets, and modified shrimp and crab pots. All take advantage of the octopus' preference for dark, confining spaces.

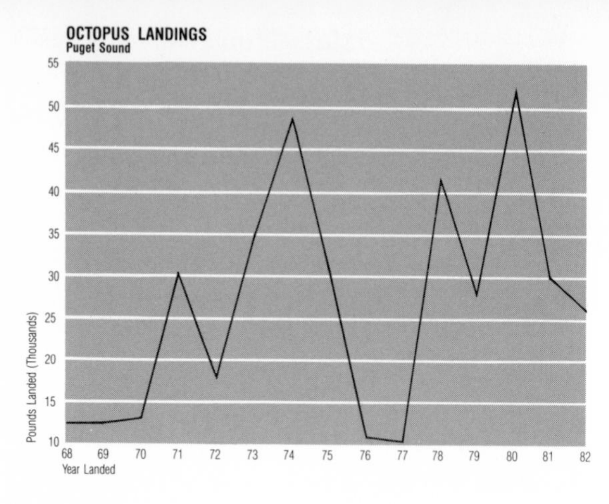


Puget Sound octopus landings (whole weight). (Source: Annual landing data for 1968–82, Washington Department of Fisheries)

## Commercial Production and Marketing

The octopus fishery is not an important commercial fishery in Puget Sound: a maximum of 52,000 pounds were landed annually from 1968–1982. Most are taken by otter trawl and fetch low prices because they are used more for bait than for human consumption. Octopus is an excellent seafood and 80 to 85 percent of the live weight is a tasty, nutritious white meat.

There is a very large market for octopus as a seafood in Japan, China, the Republic of Korea, and other Far East countries. The market in Mediterranean countries is just as great. These figures are particularly revealing—the annual harvest of octopus in Spain from 1978–1981 was as high as 196 million pounds; in Japan it has consistently exceeded 97 million pounds; in the United States, only the Washington catch is counted. Japan is also a significant importer of octopus—over 140 million pounds from Spain in 1980, in addition to imports from the Republic of Korea and other countries. An estimated five million pounds of Japanese octopus are exported to the United States for use as halibut bait.

Octopus in Spain is used as an appetizer and, among lower income groups, as a main meal. It is cheaper in Spain than most other fish or meat and about half the price of squid. In Japan octopus is regarded as a medium-grade seafood item, somewhat less desirable and costly than cuttlefish (a squid-like creature), and considerably more expensive than squid. The largest quantities are consumed fresh—often raw—in Hokkaido and western Japan, where distances from harvest area to market are relatively short. There is also a big market for octopus that is dried (*Nagaashi-dako*) and smoked (*Onkum-dako*), especially during the winter holiday season.

Octopus is still uncommon in most retail seafood markets. Many of the small to medium-sized octopuses are imported from Japan. However, increased domestic use of raw seafoods and popularity of Japanese sushi has stimulated interest in Puget Sound octopus.

## Potential for Expansion

For a larger fishery to develop in Puget Sound, an appropriate management policy and an improved market demand would be required. Octopus has been overfished in some areas of Japan despite a fast growth rate and excellent survival during the larval phase. The Japanese response has been to manage stocks by limiting the amount of gear permitted per vessel and to initiate minimum size restrictions.

The primary factor limiting population size is the number of suitable den locations for an octopus to rest on the seafloor. The octopus is a territorial animal, and fishermen report that rarely does more than one octopus occur in the same trap or pot; if two are caught in the same trap, one will usually kill and eat the other. Where suitable octopus habitat is limited, it is possible to increase the population base by constructing artificial dens and reefs.

The Washington State Department of Fisheries has conducted a test fishery in the Dungeness area and funded a modest research program to study octopus behavior. These projects were aimed at assessing alternative fishery techniques to determine the best procedure for capturing octopus, given various types of bottom conditions.

Distribution of octopus and squid harvests in Puget Sound, 1976–77. Each circle depicts average annual landings from the value given to the next highest value.

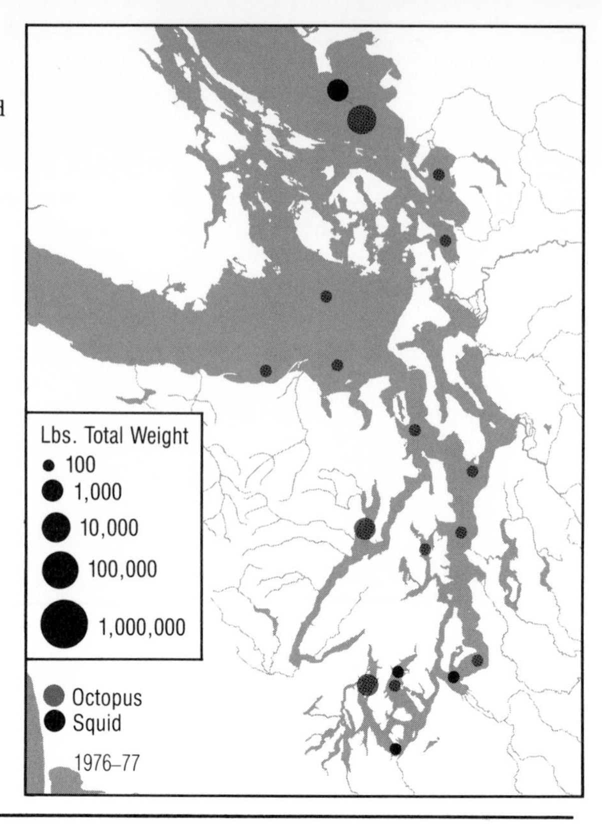


While the major markets for octopus have already been developed in countries like Spain and Japan, interest has begun to grow in the potential offered in the United States and Northern Europe. There has been somewhat of an incidental market in the Oriental and other ethnic communities for octopus; however, the Japanese market has generally been for a smaller and different species. A major problem to be overcome in large-scale marketing is the repugnance generally felt by most consumers in these markets to the physical appearance of octopus; this is also a problem for squid, and the solution is discussed later in this chapter. Although it is possible to export octopus to Japan, product size, quality, and packaging standards are very strict. Japanese importers often send their own experts to exporting countries to assist them in improving their processing, packaging, and shipping, all of which are considerations in the development of a Puget Sound fishery.

The potential for aquaculture production of octopus or enhancement of wild stocks by artificial rearing of juveniles has yet to be determined. Experimental procedures have been developed to produce and rear larvae and immature octopuses. Because the growth rate is fast and conversion of food intake to body weight is efficient, cultured octopus presents a good prospect for aquaculture. It is well to note that before release of large quantities of young or juvenile octopuses into the Puget Sound environment to increase stocks can be considered, possible conflicts with shrimp and crab fisheries would have to be explored. Such releases would have to be included in a management scheme for all competitive fisheries in the Sound.

## SQUID

The Pacific Coast squid, *Loligo opalescens*, is fairly common in Puget Sound. This striking shellfish is related to clams, oysters, and abalone, and occurs in varying numbers in the open waters of principal basins of the Sound, subtidally in inshore areas, and over sand and mud substrates when it is bottom-oriented. It grows to eight to ten inches and a maximum age of about three years. When young, it feeds upon small curstaceans and fish larvae. As an adult, it is an active predator and will even feed on other squid. It spawns in shallow water and lays its egg capsules on the seabed in late summer.

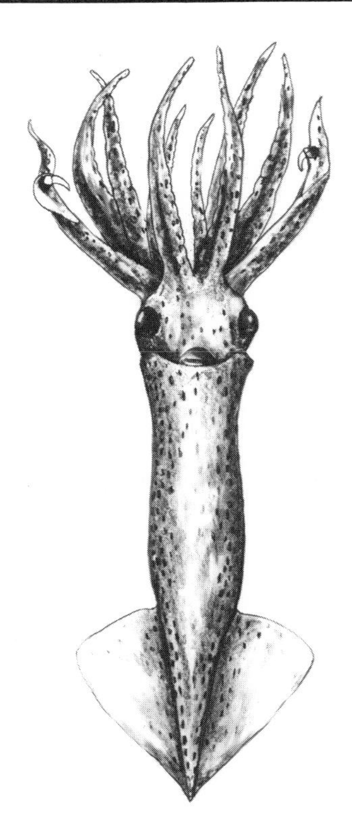

The Pacific Coast squid is the most common of several species found in Puget Sound. This is a pelagic creature with a body (excluding head and tentacles) about six inches long, with eight arms and two longer specialized tentacles.

Puget Sound squid landings (whole weights). (Source: Annual landings data for 1968–82, Washington Department of Fisheries)

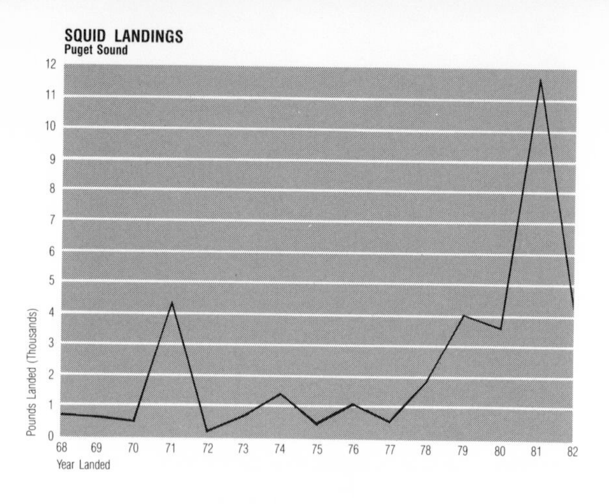


## Harvesting

Squid are harvested commercially using a variety of gear, such as jigs, trawls, lampara nets, suction pumps, and brailers. Jigging is commonly employed by Japanese fishermen in the Pacific and is popular with sport fishermen in Puget Sound. The jigs are artificial lures, and usually have barbless hooks. The Japanese have expanded jigging to a large scale and now use automatic jigging machines on the hundreds of vessels operating in the North Pacific.

Trawling is another profitable method for fishing squid, particularly on the North Atlantic. Squid, often the *Loligo* species, are taken by otter trawls towed at fairly high speeds (3–4.5 knots). Off the California coast, surround or brail nets are used at night with attractant lights. A suction pump—also known as the "squid slurp"—is also effective in California waters.

## Commercial Production

Squid are an extremely valuable fishery resource. At least 26 species, in two families, are harvested in a fishery that extends throughout the world oceans. Worldwide production of squid, along with octopus and cuttlefish, has increased 50 percent in the last decade. Squid landings now amount to about 1.4 million tons.

The Japanese are the largest producers of squid, with a catch in 1980 of almost 740,000 tons. Japanese squid fishing fleets roam worldwide, with major grounds off New Zealand, Canada, the United States, and West Africa. Other important squid producing countries include the Republic of Korea, Spain, Canada, and the U.S.S.R.

In the United States, squid has been an underutilized fish species until recently, and there is now much interest in developing a domestic squid fishery. Fishing operations along the Pacific Coast extend from

Squid are appearing in markets and restaurants as "calamari". This simple name change, coupled with greater use of fillets and other processed forms, has led to increased demand and price for squid products.

California to Alaska, and annual landings in the last decade averaged about 12,000 tons. There is increasing effort in harvesting stocks off the New England and Middle Atlantic states—where the fishery has historically been restricted to incidental catches. Total annual U.S. landings are showing an upward trend and are now about 20,000 tons.

Squid are rarely commercially taken in Puget Sound and usually then only incidentally in otter trawls. Annual commercial landings from 1968–82 have ranged from about 200 to 11,500 pounds. Recently, there has been a slight increase in total landings.

While squid has only recently entered the American market in quantity, in many parts of the world, it has long been a heavily consumed and highly regarded food. In Japan squid is considered an abundant and relatively inexpensive seafood and is very popular. Squid and cuttlefish consumption per household is over 16 pounds per year or 16 percent of total fish consumption. Squid is also popular in Europe, especially Italy and Spain, and in Korea and Thailand in the Far East and Southeast Asia.

A major problem to overcome in the U.S. market has been a strong consumer bias against the name "squid," and the traditional way of selling it whole and unprocessed, because Americans are unaccustomed to cleaning and dressing it. Calling squid by its Italian name, *calamari*, and processing it into fillet-like forms has increased consumer acceptance. Squid now appears frequently in many local markets.

## Developing a Puget Sound Squid Fishery

The absence of a well-established commercial squid fishery in Puget Sound, coupled with the competition that would occur with imports from California, suggests that the chance for success of full-scale commercial harvest in this area is tenuous at best. There is very little information on local abundance, recruitment rates, growth rates, and interactions with other species.

While it may be possible to rear squid in captivity, the technology is not sufficiently advanced to assess its commercial feasibility. Young squid require a constant supply of live crustaceans, such as planktonic animals of widely varying sizes, and they are sensitive to handling stress when held in small enclosures. Cultured squid—even in very efficient (or subsidized) systems—would be much more expensive than those caught in the wild. Therefore, there is no doubt that the fishery will continue to rely on the wild-capture harvest.

# SCALLOPS

Scallops are a distinctive and easily recognized shellfish that includes about 350 species, mostly found in warm, shallow seas—although most important commercial species are found in cool, temperate waters. There are four species of scallops found in Puget Sound, but they are not important commercially: the weathervane or giant Pacific scallop, *Patinopecten caurinus*; the spiny and pink scallops, *Chlamys hastata hericia* and *C. rubida*; and the purple-hinged rock scallop, *Hinnites multirugosus*. The weathervane scallop is the largest, growing to seven inches in diameter or more; the rock scallop reaches about six inches; and spiny and pink scallops rarely grow larger than three inches.

Except for the rock scallop, which under natural conditions is firmly attached to the bottom, these species are all bottom dwelling but capable of free swimming, though juvenile scallops—like mussels—may attach themselves periodically to a hard substrate using horny byssus threads secreted by a gland in the foot. Scallops live by filtering plankton, micro-organisms, or detritus from the water. Growth in some species can be extremely rapid—greatly exceeding that of other commercial bivalves.

## Harvesting and Processing

In the worldwide scallop harvest, the catch is dominated by three species from Japan, the U.S. East Coast, and Europe. Nearly all wild-capture fisheries use dredges, some as large as 16 feet wide. The dredges scrape the scallops—along with other bottom-living animals and rocks—off the bottom. Many scallops are able to swim away from the dredge, so the catch efficiency is usually less than 15 percent. In some areas, where the scallops are found in shallow waters, they are harvested by raking.

There is no commercial fishery in Puget Sound targeted directly to scallops, although a small commercial fishery for weathervane scallops once existed in the Strait of Georgia. Most Puget Sound scallops are now taken only incidentally by the groundfish fleet, although there is a personal-use fishery. The animals are harvested by divers using snorkels or scuba gear and a hammer and chisel to detach the shellfish from the rocky substrate.

## Commercial Production

Between 1935 and 1952 annual scallop catches in Washington State averaged approximately 96,000 pounds of meat. Between 1968 and 1982, production was erratic, ranging from no reported landings to a high of nearly 70,000 pounds in 1981. Weathervane scallop stocks have been too limited to sustain a consistent fishery, and high harvesting and processing costs have prevented the pink scallop fishery from developing. For similar reasons, there has never been a commercial fishery for rock scallops.

Two scallop species harvested commercially or suitable for aquaculture production. Top: The spiny scallop, *Chlamys hastata hericia,* is distinguished by prominent unequal ears and rasplike radial ribs. Bottom: The weathervane scallop, *Patinopecten caurinus,* has a large, nearly circular shell, equal or nearly equal ears, and a very convex left shell. Also found in Puget Sound are the rock scallop, *Hinnites multirugosus,* and the pink scallop *Chlamys rubida.*

Puget Sound scallop landings (whole weights). (Source: Annual landings for 1968–82, Washington Department of Fisheries)

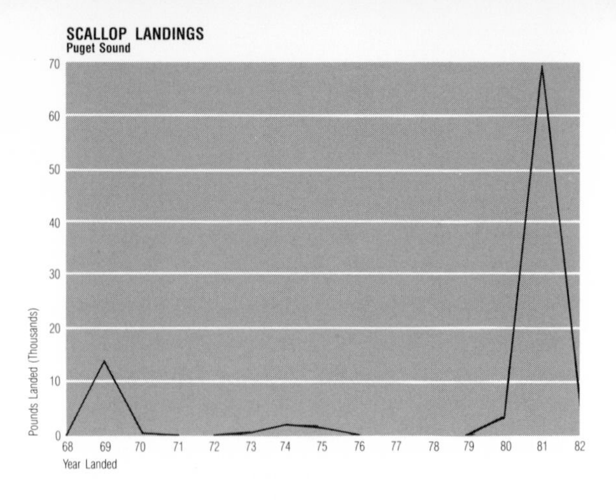


## Limitations on Commercial Harvest

Except for incidental harvest, a commercial scallop fishery is not likely to develop in Puget Sound. Significant concentrations of weathervane scallops have been fished out, and the remaining populations dispersed. The pink scallop is too small to meet the needs of the typical scallop fishery, which uses only the adductor (shell-closing) muscle, and is usually found over terrain that is too rocky for efficient harvest.

The potential for developing a scallop fishery in Puget Sound is limited, because such a fishery would require major changes in growing and harvesting methods. The existing fishery is constrained by several factors: slow recruitment of scallops, which also makes them susceptible to overharvesting; the dispersed populations of weathervane scallops; the small size of the pink scallop; and the potential for conflicts with other fisheries when dredges are used. Scallop dredges can damage nontargeted bottom organisms, such as crabs, which can cause problems when crab densities are high.

## Potential for Aquaculture

Despite the limitations, there are two possibilites for developing a scallop fishery in Puget Sound: adapting Japanese scallop culture procedures to the region; and, if a market could be developed for whole scallops or scallop meat with roe (rather than just the adductor muscle), harvesting the smaller scallops.

Limited supplies of wild-harvest scallops have led in recent years to the development of scallop aquaculture. Production of cultured scallops in Japan now greatly exceeds harvest of natural stocks. Scallop seed is collected by placing coarse mesh bags (often plastic onion sacks) filled with used nylon netting in the water just before the larvae are ready to attach. The seed scallops are suspended below rafts or floats, or hung by the "wings" on the shell (called "ear hanging"). Scal-

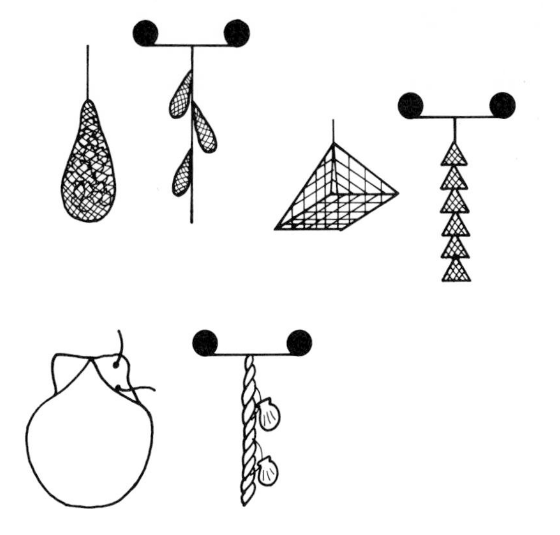

Details of longline landing culture methods for scallops (longline floats are shown as black circles). Above left: Seed collecting sacks two-feet long are suspended from a 25-foot line; 15 sacks can be suspended from a single line. Above right: Pearl nets are small pyramidal suspended baskets used to rear scallops to an intermediate size. Bottom: Ear hanging, one of several methods that can be used to grow out harvestable scallops in hanging culture.

lops grown under these conditions have lower mortality than natural stocks, and they reach market size in 18 months or less.

The Washington Department of Fisheries has investigated the potential of establishing a scallop farm in Puget Sound to enhance and replenish natural stocks and provide seed for commercial growers. The most encouraging results were obtained with the rock scallop, a species which also showed promise in California. Studies suggest that the rock scallop may be well suited to aquaculture because it normally attaches to a stationary substrate. Young rock scallops held in cages in plankton-rich waters grow about two and a half inches per year, have good meat yields, and don't have the massive shells of older ones. These studies are still experimental, and much more work is needed to confirm the commercial potential for rock scallop culture.

Scallops are a valuable and expensive fishery product. They are accepted by consumers worldwide. Whole scallops are marketed in Japan; only the adductor muscle and gonads are marketed in Europe; and only the adductor muscle is marketed in the United States. Because in most traditional markets the adductor muscle only is marketed, the small pink and spiny scallops are at a competitive disadvantage to the large East Coast and weathervane scallops. The potential for developing a fishery for smaller scallops may depend on selling the whole animal, for example as oysters are sold to the half-shell trade. Whole scallops have an excellent flavor and texture, but they do not stay fresh as long as oysters and can retain PSP toxins. When exposed, scallops ingest and retain high levels of toxin, thus—as is true with oysters, clams and mussels—PSP outbreaks can result in closure of commercial scallop harvests.

The pinto abalone, *Haliotis kamtschatkana*, is a native species, commonly found in northern Puget Sound. Its shell reaches a length of 4½ inches and has a rough or wavy surface. All abalone shells are distinguished by a row of holes paralleling the outer shell margin.

# ABALONE

Abalone are one of a small group of marine snails with a highly profitable commercial market. Commercial-sized abalone occur in the northern and southern temperate and subtropical zones. Principal landings are from Pacific Ocean waters along Japan, Australia, and North America. Two species are found in Puget Sound: the native pinto abalone, *Haliotis kamtschatkana*, and the red abalone, *Haliotis rufescens*, which was introduced from California in 1958. There is no commercial fishery allowed for either of these species in Washington waters, but they are harvested commercially in Alaska and British Columbia. Pinto abalone are allowed to be taken in the recreational fishery, and are fished in small numbers by local sport divers.

Abalone occur naturally in rocky areas exposed to oceanic waters. Young, juvenile abalone less than 1/2-inch long feed on small plants (diatoms), and remain hidden from view in cracks and crevices. As the abalone grow, they feed on larger algae or macrophytes and begin to move about on more exposed rocks, although their range of movement generally remains very limited.

Pinto abalone are markedly smaller than the other West Coast species. The shell measures up to six inches long, but most are less than four inches. The red abalone, on the other hand, grows to 11 inches or more and attains a weight of up to 5 1/2 pounds (total weight). There is no documented evidence of natural reproduction of red abalone in Puget Sound; however, recently sport divers have harvested some immature but large abalone from the San Juan Islands. This suggests that red abalone are reproducing and may be hybridizing with the pintos.

## Harvesting

Almost all of the world abalone catch is taken by divers who remove the animal from rock with a curved bar or chisel. Despite harvesting difficulties, abalone are often overfished, and management actions are required to conserve the resource. Such actions include limitation or reduction in total catch, minimum size limits, rotation of fishing grounds, and seasonal or area closures. Despite these measures most of world abalone fisheries are experiencing declining catches, reduced stock densities, and lower average sizes. This indicates that recruitment of young abalone is not adequate to sustain the population. For this reason, government agencies and private operations in the United States are attempting to rear abalone seed and are testing the potential of restocking abalone beds. The methods used are based on a Japanese model, which now accounts for 44 hatcheries producing seed for the fishery.

## Production and Marketing

There is no commercial abalone production in Puget Sound. Landings (1981 annual catch figures) along the West Coast of North America are dominated by Mexico (about 1,900 tons) and the United States (about 800 tons). During the same period the landings in British Columbia were just over 81 tons. Abalone landings along the entire West Coast have been declining in recent years, and present harvest levels are less than 20 percent of the harvests of 10 to 15 years ago. Abalone production in Japan (1970–79) has averaged about 6,000 tons annually and is presently enhanced by massive reseeding programs of several billion seeds per year. Australian landings are about the same as Japan's.

Abalone is one of the most expensive fishery products sold in the United States. Typical retail prices for abalone in Seattle are $10 per pound for frozen meats. The very high prices are due primarily to the scarcity of abalone stocks in U.S. waters and the high cost of harvesting.

The highest quality abalone is cleaned and sold fresh or frozen, with the biggest markets in Japan and the United States. A harvestable abalone (four to five inches) weighs slightly less than a half pound and contains about 40 percent edible meat. Abalones that are small or of a lower quality are generally smoked, or boiled and canned. There is also a market for dried abalone in Asia and Hawaii.

## Establishing an Abalone Fishery in Puget Sound

Since the abalone is not a commercial species in Puget Sound, this discussion is limited to points relevant to the establishment of a potential fishery in these waters based on hatchery-reared seed. Suitable sites for abalone culture are limited to areas of natural distribution—in the northern portions of Puget Sound, the San Juan Islands, and the Strait of Juan de Fuca. Abalone cannot be grown in south Puget Sound. Although they adapt to turbulent water conditions, they can be dislodged by heavy wave activity and are subject to predation by fish (the most significant natural predator), sea otters, starfish, and octopus. Abalone are also affected by pollutants and can accumulate lead and mercury.

Because it is unlikely that natural production could sustain a commercial harvest in Puget Sound, there has been considerable interest in augmenting natural stocks and in commercial abalone rearing. The Washington Department of Fisheries, adapting techniques developed in California and Japan to produce pinto and red abalones, developed a pilot hatchery and studied culture conditions, methods, and techniques. This work was envisioned as a means to replenish natural stocks, and it was hoped that production could be increased so that some seed could be sold to future private growers. The program successfully reared young abalone, but overall results were not encouraging. The long growth period of pinto abalone—coupled with the unknown mortality due to transplanting and high costs—made that species a poor candidate for commercial success. Similar problems occurred with red abalone culture, and there was additional concern over the impact of large-scale plantings of a non-native species.

These preliminary experiments could, when augmented by studies of methods for increased production, indeed form the basis of enhanced natural production and restocking wild abalone populations in Puget Sound.

## OTHER MOLLUSCS WITH COMMERCIAL POTENTIAL

The species discussed so far by no means exhaust the marine organisms suitable for human consumption. Other little known molluscs have the potential to be harvested commercially and perhaps cultured in the Pacific Northwest. In Hawaii, limpets (known as *opihi*) are coming under increasingly heavy fishing pressure. The price for shucked fresh limpets ranges from $75 to $150 per gallon in the local Hawaiian retail market. The possibility of a limited local harvest of limpets could be explored as a means of augmenting Hawaiian stocks. Unfortunately, the price is likely to be lower because Hawaiians believe Pacific Northwest limpets are not nearly as tasty as the tropical *opihi*.

Moon snails are common predators of clams on intertidal beaches and are removed as pests by clam farmers. One grower noted that he was able to sell moon snails to Japan, although the supply and market demand was highly variable. Moon snails are excellent to eat, either steamed or fried, and were at one time eaten by the Indians who lived along Puget Sound. Commercial use of moon snails where they are abundant could partly compensate for the economic losses that they inflict upon clam farmers.

The commercial exploitation of limpets, moon snails, and other intertidal and subtidal molluscs would have to be carefully evaluated and monitored to minimize any ecological damage or impact on other

Other molluscan shellfish, such as chitons, limpets, and moon snails are edible, particularly when used in well-crafted recipes. However, lack of a commercial market, and unknown or inadequate supplies preclude their consideration for a commercial fishery.

harvested species. It is not likely that a platter of littorines, chitons, or fusitritons will be used soon as an evening meal, but the potential of these species to meet future demands of more sophisticated seafood consumers or export markets should not be discounted.

The early years of the Dungeness crab fishery were dominated by the use of baited pots or traps and ringnets or hoopnets. Additional numbers of crabs were also taken by gill and trammel nets. (Photo by Asahel Curtis, courtesy Special Collections Division, University of Washington Libraries, negative 13297)

CHAPTER FIVE

# Crabs

Harvesting of crabs has had a long history in Puget Sound and still is a major shellfishery of these waters. In the mid-1800s, early settlers began catching crabs along the shore from the intertidal zone of nearby bays and inlets. Later, they expanded their harvest areas by devising nets and traps that could be fished from small boats in shallow waters. By 1920 crab harvests in most areas of Puget Sound were about as high as the stocks could sustain, and crabbers began to look to the nearshore waters of the open coast for new hunting grounds. The greatest number of crabs presently landed in Washington State are from coastal waters, yet Puget Sound sustains a significant fishery.

The crab fishing industry in Washington State is based on a single species, the Dungeness crab (*Cancer magister*). It has been called at vari-

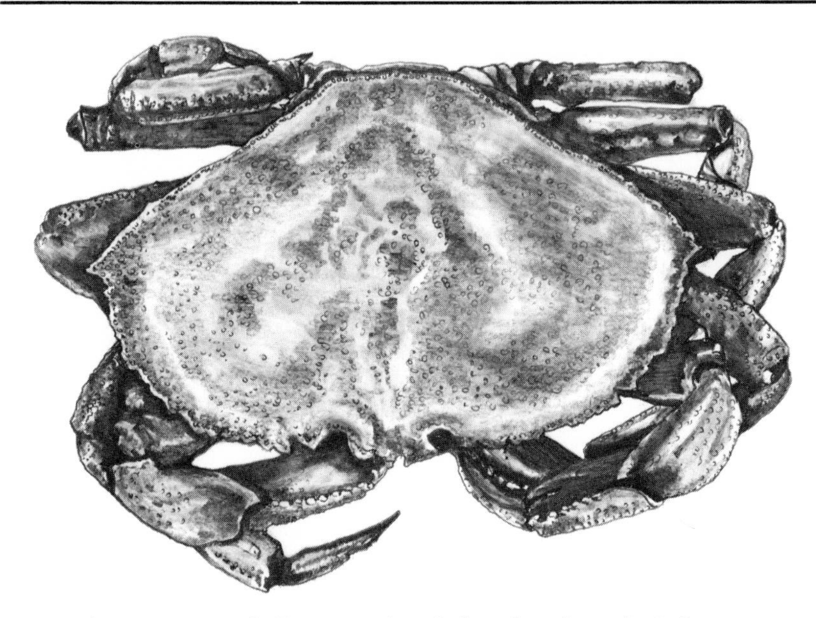

The Dungeness crab, *Cancer magister*, is the only crab species in Puget Sound harvested in large numbers in commercial and sport fisheries. It grows to almost nine inches across the width of the carapace.

ous times the market crab, commercial crab, edible crab, or Pacific edible crab—names that were a testimonial to its utility. It is found from Alaska's Aleutian Islands to Southern California and throughout Puget Sound, inhabiting the low intertidal to deep subtidal zones. In Puget Sound it is most abundant in the northern Sound in subtidal waters up to 400 feet in depth, where the substrate is sandy or muddy.

Dungeness crabs do not migrate over long distances or between major fishing areas, but rather move in and out of the same bay or occasionally from one bay to the next. Although some tagged crabs have been recovered more than 100 miles from where they were released, most are recovered within a few miles of their release point. They feed on clams, small fish, other crabs, and marine worms.

Dungeness crabs are polygynous animals (males mate with more than one female), and for this reason only male crabs are allowed to be harvested. They mate in shallow water, usually in late spring or early summer, immediately after the female molts but before her new shell hardens. She stores the sperm in a seminal receptacle. The eggs are fertilized in the fall, when they are extruded and become attached to the abdomen. Large females may carry up to 2.5 million eggs. Egg-bearing females can be found year round, but they are most common from November through February. In Puget Sound, eggs hatch from March to May.

In the larval stage, when newly hatched crab larvae join the free-floating plankton, many are eaten by shrimps, herring, young salmon, and

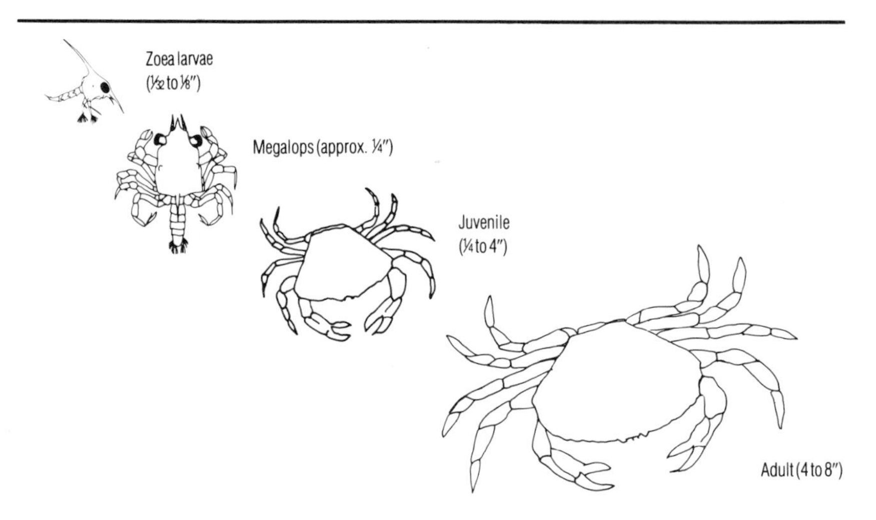


Life cycle of the Dungeness crab. The planktonic (free swimming) larvae go through six phases—five as zoea. Each stage is accompanied by a molt (loss of shell) and increased size. Juveniles grow through about 12 molts before becoming sexually mature adults.

other marine predators. Those that survive pass through six larval stages, swimming free for three to five months before settling on the bottom to begin life among the benthos. As the young crab grows, its progress is marked by a physiological process called a molt, which occurs in all crustaceans with external skeletons. A crab may molt as many as six times during its first year; twelve times before it reaches sexual maturity; and up to fourteen times before it reaches the minimum legal commercial size of 6¼ inches across the back when it is about four years old. Once the carapace is about four inches wide, the crab grows about one inch per molt. Males and females both grow at about the same rate until they reach sexual maturity; then females grow more slowly than males, which can attain larger maximum size (about 8⅝ inches).

## Crabs and the Environment

Dungeness crab populations have very high mortality rates due to natural causes and the effects of fishing. Natural mortality caused by environmental factors varies depending on molt stage, condition of the animal, and the development phase of the crabs. It is especially high among larvae, which are eaten by plankton-feeding animals, and among small juvenile crabs, which are consumed by fish and octopus.

Dungeness crabs prefer cool waters, and variations in temperature, salinity, and oxygen concentration can adversely affect their health. Increases in temperature can slow growth, lower feeding activity, reduce hatching success, and cause death, especially when the crabs approach molt. They have limited ability to control inner water content if the salinity of the water decreases. A decrease in dissolved oxygen concentration lowers their resistance to stress caused by changes in temperature or salinity.

Because Dungeness crabs are bottom-dwelling creatures, changes in substrate also affect their well-being. Modification of habitat in shallow-water estuaries disturbs important nursery areas for juvenile crabs. Adult crabs forage over tidal flats, ship channels, and secondary channels, where dredging and siltation can crush or smother them and their prey. They are especially susceptible to this type of habitat modification during their mating season.

Pesticides and insecticides washed by rivers and storm drains into marine waters also threaten crabs. Even small concentrations (2–4 parts per million) of some chemicals are toxic, depressing reproductive functions and lowering resistance to other stresses. Crabs are particularly sensitive to organophosphates, carbamates, and chlorinated hydrocarbons, which can kill them as effectively as they kill mosquitos. The Dungeness crab fishery in the San Francisco area is affected by the discharge of pollutants—such as grease, oil, heavy metals, and pesticides—into San Francisco Bay, the principal nursery area for crabs. This is a lesson that should not be ignored in Puget Sound.

Crabs trapped in pots endure different stresses. First, they must contend with each other in confined quarters; mortalities in traps are usually attributed to cannibalism. These mortalities are significantly higher among crabs caught while molting, since newly molted crabs are especially vulnerable to those that still have hard shells. Crabs stored in live tanks aboard ship or at dockside holding facilities are subject to additional physical and chemical stress. Accumulated ammonia, the chief excretory product of crabs, is toxic at low concentrations. Further, although few diseases have been documented among crabs in the sea, a number of bacterial and fungal infections occur among crabs held in high-density, stressful conditions. These diseases are fairly common, yet their impact on survival of natural free-roaming stocks is unknown. Predation from a wide variety of animals accounts for substantial crab mortalities. One worm-like predator commonly infests brooding females, where it feeds off the egg mass. Coho and chinook salmon are important predators—up to 1,500 crab larvae have been found in the stomach of a single young salmon.

Competition within crab populations may also play an important role in crab survival. When crabs are very abundant, there are large numbers of adult females and fewer males after harvesting to compete with the juveniles for food. Because nature tends to favor larger, stronger animals there are likely to be higher mortalities among the young crabs, causing fluctuations in abundance in future years.

## Harvesting

In Puget Sound, Dungeness crab are fished along tidal flats and in shallow waters where crabs bury themselves in the sand, seaweed, and debris. Although anyone can harvest crabs by hand for his own use, commercial crabbers are only allowed to use ring nets and pots.

Ring nets were common in the early commercial fishery, and are still used by crabbers today. They are constructed of one or two wooden or metal hoops that have a mesh-covered opening. A bridle and line are attached to the net so that it can be lifted flat from the water. Bait (fresh or frozen clams, squid, or herring) is tied to the center of the net, which is lowered to the bottom and lifted periodically to remove the crabs. More effective ring nets are made by adding a second ring and extra mesh to provide side walls that create a bucket shape.

Most of the Dungeness crab catch is taken in baited pots. Although there have been many designs devised for crab pots, most modern commercial pots are drum-shaped, about 36 to 49 inches in diameter and 18 inches high. The frame is constructed of iron rod, which may be coated with rubber to decrease electrolysis, and covered with two- to three-

The modern Dungeness crab fishery differs little from that of the early 1900s. Use of the hydraulic line puller (above left), and pots made partly of manmade materials (left) have increased the efficiency and flexibility of the fishery. Smaller-sized crabs are measured by hand (above right). (Photos courtesy Washington Sea Grant Marine Advisory Services)

inch (five- to eight-centimeter) stainless steel mesh. There are two entrance funnels leading into an opening (also called the "eye") of the pot, which is equipped with stainless steel triggers to prevent the crabs from escaping, and a hinged lid for removing them. In addition, all commercial crab pots must have two 4¼-inch escape rings to allow females and undersized crabs to escape.

Crab pots are fished singly or in strings, each pot equipped with a bridle, line, and buoy branded with the owner's identification. The Washington Department of Fisheries regulates the number of pots that commercial vessels can fish, up to 100 pots per boat except in Dungeness Bay where the limit is 60. They are usually set in lines, 5 to 20 fathoms deep, at roughly equal intervals along a chosen course. Pots are generally fished from one- to ten-day periods, depending on the time of year, weather and sea conditions, and abundance of crabs. Boats sometimes work several strings set parallel with each other so that they can be located and recovered even in rough or foggy weather.

When a pot is brought aboard, the catch is removed and legal-sized crabs sorted to be held on deck or put into a recirculating seawater tank. The rest of the catch, any fish and undersized and female crabs, are returned to the sea, the bait is replenished, and the pots reset. An efficient crew can pick and reset all their pots in a single day and may fish up to 24 hours straight through. One crewmember hauls the pot aboard with a hydraulic power block, while another fills bait cans or sorts the catch. The boat remains underway during the entire picking and setting operation, maintaining a speed of about two knots.

## Marketing and Production

Dungeness crab is marketed in two forms, either cooked in the shell or as cooked shucked meat, which yields 25 to 35 percent of live weight. Whole crabs are cooked immediately upon arrival dockside, and are marketed fresh or fresh-frozen. Most shucked meat is packed in one- or five-pound containers and is also marketed fresh or fresh-frozen, although some shucked meat is heat processed and canned for the retail trade. A market has also developed for live crab, which are shipped by air to Hawaii and other distant points.

Market price and demand for Dungeness crabs vary seasonally. Prices are generally lower at the beginning of the commercial harvest season, when stocks are abundant and catch rates high. Later in the sea-

Dungeness crabs—here lined up for sale at Seattle's Pike Place Market—are very popular in local retail markets and offer an attractive contrast to other shellfish and finfish.

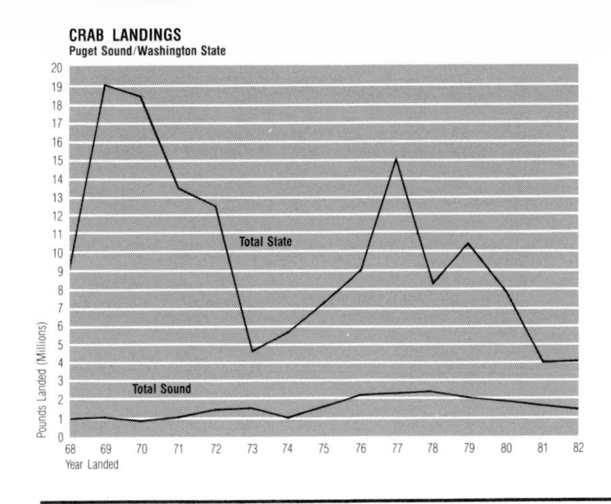


Washington State and Puget Sound Dungeness crab landings (whole weights). Recent landings in Puget Sound have accounted for nearly half of all state production. (Source: Annual landings data for 1968–82, Washington Department of Fisheries)

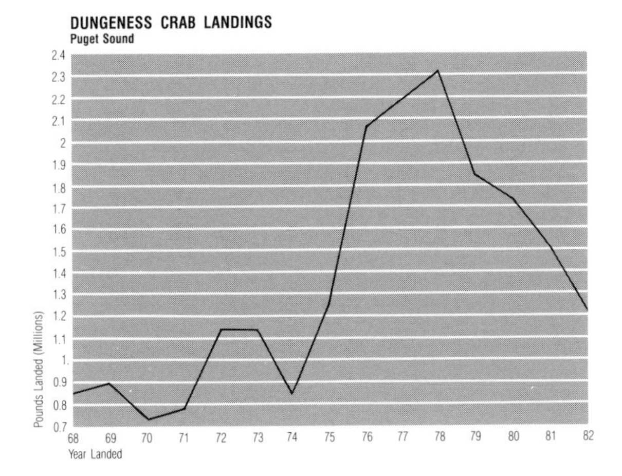


Puget Sound Dungeness crab landings (whole weights). (Source: Annual landings for 1968–82, Washington Department of Fisheries)

son, as crabs become scarce and catches fall off, the price increases.

Dungeness crab constitutes an important fishery in Puget Sound and total landings now rank fourth behind geoduck, oysters, and hard-shell clams. From 1952 to 1976, the annual commercial harvest of Dungeness crab in Puget Sound averaged slightly over one million pounds, or about 11 percent of the total state catch. In 1975, production began to increase and from 1976 to 1982 annual landings averaged 1.8 million pounds, causing Puget Sound's share of the statewide crab harvest to rise to about 30 percent. Puget Sound Dungeness crab stocks were quite stable from 1974 through 1984. Although some fluctuations in abundance occurred, they were not of the magnitude of the "boom or bust" levels reflected by the coastal crab fisheries. Worldwide, in 1981 Dungeness crab accounted for 39 million pounds of a total catch of 1.9 billion pounds for all crab species taken.

## Crab Resource Management

Management of the Dungeness crab fishery is based on data derived from fishing effort, stock assessment, mortality, and cycles of crab abundance. In Puget Sound, regulations are applied to gear type and quantity, and duration and timing of fishing seasons.

Considerable effort has been expended to determine how to manage the Dungeness crab fishery considering molting cycles, reproduction, and the various factors affecting mortality. Up to 60 days after crabs have molted, meat yield is low—only 13 to 14 percent of live weight. Also, the meat tends to be watery, and the soft-shelled crabs are sensitive to handling. Because hard-shelled crabs have much higher meat yield and are hardier, an attempt is made to restrict fishing to periods when the crabs are not molting. This is difficult to accomplish, however, because crabs do not always molt at the same time each year or at the same time throughout their range.

Each year, an estimated 10 to 15 percent of all pots used are lost to the seafloor, where they continue to trap crabs—sometimes for up to four years—which die and then become bait for other crabs. To prevent these needless mortalities and capture of undersized crabs and females, management agencies are requiring or testing various types of "savings gear," mechanisms such as escape rings and destruct devices. Escape rings are designed to allow small crabs to get out of the pot, thereby reducing mortality of undersized crabs due to in-trap cannibalism and unnecessary handling. Escape rings also reduce the time required for fishermen to sort their catch. Destruct devices cause the pot to open, permitting crabs to escape after a period of time. New materials are also being developed that would cause part of the closure mechanism to disintegrate over time, thereby releasing the lid so that crabs could escape. Use of destruct devices is not yet required.

Regulatory agencies also attempt to alleviate interfishery conflicts among crabbers, trawlers, and salmon gillnetters. When crab pots are set in trawl or gillnet areas, nets can become damaged and fouled by pot lines and buoys, and crab pots lost as the fouled lines are cut. In addition, large numbers of crabs are caught incidental to bottom trawl and net fisheries, which cause substantial damage to crab stocks.

Conflict also exists between the commercial and sport crab fisheries. This was especially a problem in the early 1970s, when a threefold increase in commercial gear resulted in opening of new areas to commercial fishing that had previously been restricted to personal use fishery.

Some of the problems may have resulted from an apparent lack of awareness by many sport crab fishermen of the biology of the crabs and of the need to regulate the fishery. A survey of intertidal sport crabbers near Mission Beach north of Port Gardner, found few who were aware

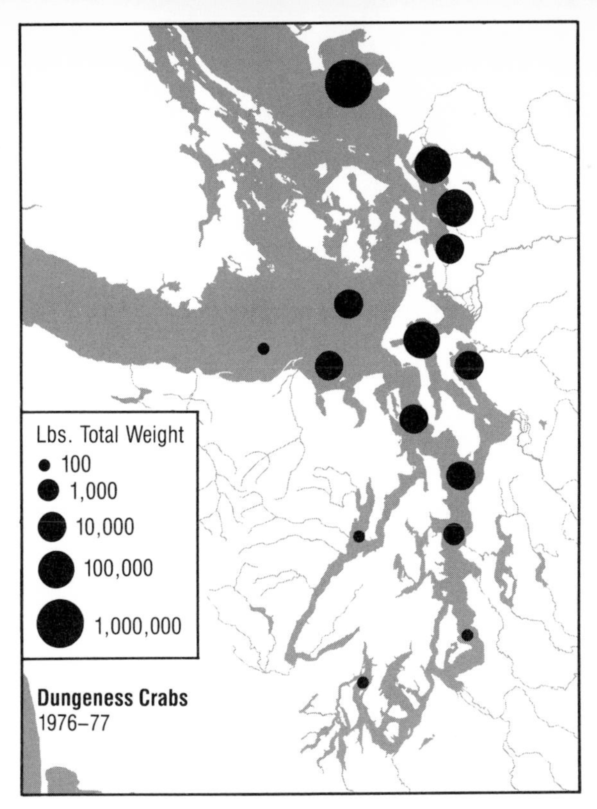


Distribution of Dungeness crab captures in Puget Sound, 1976–77. Each circle depicts average annual landings from the value given to the next highest value.

that crabs molt. They believed that the exoskeletons found on beaches were the result of commercial crabbers killing small crabs and overfishing the stocks. Sport crabbers also believed that commercial crabbers were taking most of the large crabs. A reduction in sport size limit from 6¼ inches to 6 inches did not ameliorate this conflict. Commercial crabbers countered—with some justification—that sport fishermen now harvest the 6- to 6¼-inch crabs that would normally enter the fishery after their next molt.

Sport crabbers have traditionally been concerned that commercial crabbers were taking all the available crabs. Commercial effort in Puget Sound increased during the 1970s from about 150 vessels in 1972 to nearly 400 by 1980. A license moratorium was enacted in 1980, which reduced the number of crab licenses to less than 300 (sale or transfer of licenses is prohibited). The target level established for the commercial crab fishery is 200 eligible vessels.

## The Future of Commercial Crab Fisheries

With the exception of certain areas in Alaska, known West Coast crab stocks are exploited near maximum sustainable yield, and catch rates have levelled off and are not expected to change. There is a well-established demand for Dungeness crab, and it brings a high price in the marketplace.

Stable demand and high prices might appear to make crabs an economically viable candidate for aquaculture development. Indeed, laboratory trials conducted in Oregon have shown that Dungeness crabs can be bred and cultured in captivity through the free-swimming larval stage. Yet the commercial feasibility of aquaculture of Dungeness crabs still faces many problems. Methods similar to those used for culturing Maine lobsters have been proposed, but none have been developed. Crabs would require expensive supplemental feeding, and feed conversion rates are likely to be poor. Growth rates are slow; two to three years may be required to produce a commercial-sized crab. Finally, the Dungeness crab's cannibalistic nature would necessitate special enclosures to separate and protect the soft-shelled, newly molted crabs.

The red or rock crab, *Cancer productus*, is an abundant intertidal species that deserves mention because it is sometimes caught in crab pots along with the Dungeness. It is usually found in habitat that is rockier than that preferred by the Dungeness, however, and it is not commercially valuable because of its small size. Experimental harvest permits were issued in the early 1980s but no fishery developed. Some potential may exist for such a fishery, but it is limited by low market value and gear restrictions needed to protect the valuable Dungeness resource. Another species, the black-clawed crab, *Lophopanopeus bellus*, is believed by oystermen to be a useful predator that appears to feed preferentially on oyster drills. This heavy-clawed crab has reportedly reduced the incidence of drill infestation on low intertidal and subtidal oyster beds in some areas.

The future of the Puget Sound crab fishery will remain bright as long as critical habitats are protected from damage resulting from marine industry, pollutants, and construction activities, and as long as mortalities associated with bottom trawl and salmon net fisheries are minimized.

CHAPTER SIX

# Shrimp and Other Nonmolluscan Shellfish

Man has made use of many edible foods from the sea; some of them—such as shrimp—have become common as seafoods, others—such as sea urchins and sea cucumbers—are still considered rather exotic by many consumers. Many of these marine species are found in Puget Sound. Shrimp were commercially harvested here as early as 1888 by fishermen using beach seines near Anderson Island. They soon replaced their shorebound gear with beam trawl nets that could be towed behind small steamboats, and opened the way for rapid development of the shrimp fishery. The beam trawl could easily be handled by one man from a small boat, and the light simple gear allowed many people to enter the shrimp fishery near major market centers in Seattle and Tacoma.

Shrimp grounds in central and southern Puget Sound near Seattle and Tacoma were quickly depleted, and by 1903 the fishery had to expand to the San Juan Islands and Hood Canal. There, and elsewhere in the Sound, shrimpers still work the waters with beam trawl nets or pots reminiscent of the early years of the fishery.

The more exotic species had not gone unnoticed, either. In the early 1900s, a naturalist reported on the possibility of harvesting Puget Sound echinoderms—particularly sea urchins and starfish—for food. There was no rush of enterprising entrepreneurs willing to develop markets for these creatures, however, and they were virtually ignored for the next seventy years. Today, there is significant demand for exotic seafoods in the Orient, Europe, and increasingly in the United States. Animals such as sea urchins, sea cucumbers, and jellyfish are used for food and as flavors and condiments. In Puget Sound, the globular sea urchin and its relative, the slug-shaped sea cucumber, have been the prime elements in the development of new commercial fisheries.

## SHRIMP

West Coast shrimpers catch almost exclusively species of the family Pandalidae, which concentrate in cool, temperate, and subarctic waters. Of the twelve species of pandalid shrimp represented in the world's fisheries, seven are common to Puget Sound. Worldwide, pandalid shrimp account for only a small portion of total shrimp landings,

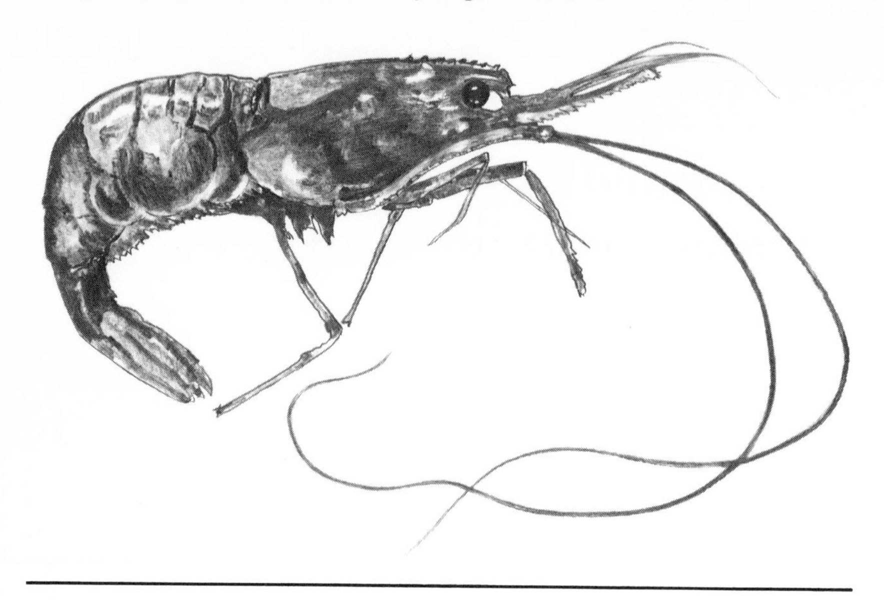

Pandalid shrimp (prawns) are the only species of commercial importance in Puget Sound. These shrimp reach a body length of up to nine inches.

about 8 to 20 percent. The bulk of world shrimp catches belongs to the family Penaeidae, which includes the Gulf Coast prawn.

Many of the same physical and biological factors that affect Dungeness crab also affect shrimp. Like crabs, adult pandalid shrimp prefer low water temperature and can tolerate moderate fluctuations in salinity. However, low salinity levels can inhibit larval pandalid growth, an effect opposite that found for many penaeid shrimps that use estuaries as nursery areas. Light is also an important growth factor. Larvae are often found to be attracted to light, whereas juveniles and adults tend to avoid lighted areas. This leads to vertical migrations as shrimp follow food stocks that are light-controlled. Adult shrimp follow their prey up from the bottom during the late afternoon and evening, and return to the bottom at dawn.

Pandalid shrimp also migrate horizontally, between nearshore and offshore waters. This migration pattern may be associated with changes in temperature or salinity, and with the onset of breeding. Pandalids are carnivores (only rarely have they been found to eat plant matter) that eat marine worms, small crustaceans (such as copepods and euphausiid shrimp), or they will feed on detritus (dead animal and plant matter).

Pandalid shrimp are protandric hermaphrodites—that is, each individual shrimp matures first as a male, then undergoes a sex change to female. Hermaphroditism is not uncommon among marine creatures: it occurs in other shrimplike species and in many molluscs. Shrimps usually spawn during their first 12 to 18 months, when they are males, and between one to two seasons later change to females, which they remain the rest of their lives. In a few cases, the female phase may develop early and the male phase may never mature.

Mating occurs in the fall. The male attaches its sperm to a receptacle on the underside of the female. She spawns 2,000 to 3,500 eggs that are then fertilized by the waiting sperm and remain in the egg sac until they hatch. In the late winter or early spring, the newly hatched larvae join the communities of other plankton and planktonic larvae for two to three months. They will pass through five or six larval stages before becoming juveniles and settling down to a semibenthic existence.

Little is known of the diseases and parasites of pandalid shrimp. A condition known as "black spot gill disease" occurs in pink shrimp off Maine, and has been observed in pink shrimp off Kodiak Island, Alaska, but it is not known whether this disease is fatal. Many species of pandalid shrimp endure parasites—the most common are small crustaceans called isopods—but the effects do not appear to present a major threat to shrimp stocks or their commercial value.

Predators, which include large predatory fish such as rockfish and dogfish shark, are the primary cause of shrimp mortality. Seasonal variations in abundance occur apparently because of high mortalities during the larval and early adult stages. If large numbers of young shrimp survive these natural causes of mortality, then it is likely there will be good stocks for a fishery.

## Harvesting

Puget Sound shrimp fishermen use either beam trawls or pots to harvest shrimp. The beam trawl, and a modified version of it called a pipe trawl, has been used locally since the fishery began and is still used in north Puget Sound and the San Juan Islands. (The type of otter trawl or shrimp trawl used in the Gulf of Mexico is illegal in the Sound.) The beam trawl is a cone-shaped net with the mouth held open horizontally by a beam running across the top and directed by skids at each end of the beam. Because it is towed through the water above the bottom, it can be fished over rocky substrate.

Shrimp boats are small (less than 60 feet in length), two-man vessels that carry only basic equipment required to winch the trawl in and out and lift it onboard. Trawl nets are sized to suit the capabilities of the vessel and crew and are usually less than 40 feet wide at the beam. The net is small enough to be towed by a small boat, but is ineffective in covering large areas.

Shrimp pots are box frames covered with wire mesh or fiber netting, with two or more tunnels sloped inward and a 2¾- to 3-inch circu-

The shrimp pot (below) and beam trawl (left) are the two methods commonly used for the Puget Sound shrimp fishery. Both methods have been practiced since the early 1900s.

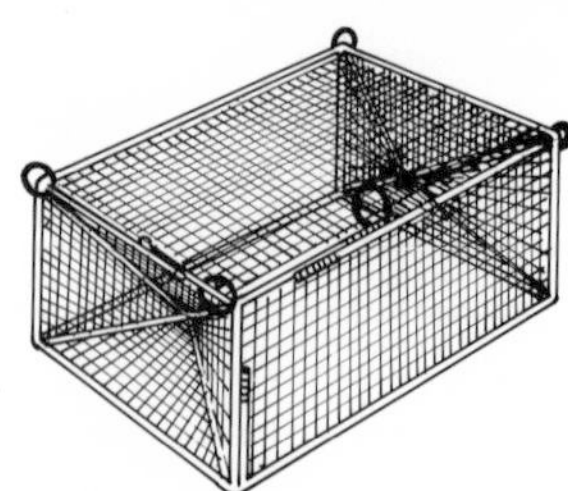

lar opening at the inner end. There is a minimum mesh size for the pots in Hood Canal and no more than 100 per boat are allowed in Puget Sound, except in Hood Canal where the limit is 50, and in Port Angeles where the limit is 10. Pots, set singly or attached by groundlines, are set at regularly spaced intervals, and are baited with such temptations as canned fish, flavored cat food, fish carcasses, and clams. A variety of small power boats, such as open runabouts (16–35 feet) and gillnetters, are used in the fishery, and the only necessary equipment is a line hauler or reel. The one- or two-person crew unloads the catch in the late afternoon, so that it can be iced for delivery on the same day and sold fresh in local markets.

The shrimp season is April 15 through October 15 for trawling in Puget Sound, and May 15 through September 15 for pot fishing. Two areas of the inland waters have special restrictions on shrimp fishing: the season in Hood Canal and Carr Inlet can be closed whenever stock abundance is estimated to be too low to permit commercial harvesting.

## Shrimp Production and Marketing

During the early years of the Puget Sound shrimp fishery, production barely kept pace with local demand. Starting from a modest catch of 5,000 pounds in 1888, annual catches grew to a peak of 400,000 pounds in the years 1903 to 1915. Catch levels in the current fishery have remained at a lower level, which has been relatively steady in recent years. Between 1936 and 1982 total shrimp production in Puget Sound ranged from a low of 9,000 pounds to a high of 150,000 pounds.

Historically, about one-half to two-thirds of the total Puget Sound catch has come from the pot fisheries of Hood Canal and Carr Inlet. Since 1980, the bulk of the pot catch has occurred in Hood Canal, where commercial and sport catches combined have ranged from 160,000–200,000 pounds annually. The beam trawl fishery, which is banned in Hood Canal, has been concentrated around the San Juan Islands and in north Puget Sound, with catches consisting primarily of pink and coonstripe shrimps and spot prawns. Spot prawns are the predominant species taken in Hood Canal. During the 1950s and 1960s some shrimp were landed from Elliott Bay, but no landings have been reported in recent years.

The pandalid catch of Puget Sound is a small fraction of all West Coast shrimp landings, which are dominated by the northern pink, *Pandalus borealis*, and the ocean pink, *P. jordani*. These slow growing and tiny shrimp (the ocean caught species average 200–400 whole shrimp per pound) were caught in huge numbers in the late 1970s, with catches of over 150 million pounds. Since then, the catches declined to just 21 million pounds in 1983, due in part to overfishing and the effects of *El Niño*. The catch off the coast of Washington has amounted to about five to ten percent of the total West Coast catch, ranging from 0.6 to 12.7 million pounds annually from 1968 through 1982. The larger shrimp taken in the Sound are also harvested off the coasts of western Canada and Alaska: the Alaska pot fishery for spot and coonstripe alone took more than a half million pounds in 1983.

The processing and product forms of pandalid shrimp are not complex. Most of the small pink shrimp are processed peeled, cooked, and frozen; whereas the larger shrimp and prawns are often sold fresh, either peeled or unpeeled—the best product forms for these fine shellfish. The smaller shrimp have proven difficult to market at a reasonable price in the face of stiff competition from imports harvested in the North Atlantic. The market for high quality large shrimp is very good with especially high prices for fresh products. Because of the limited

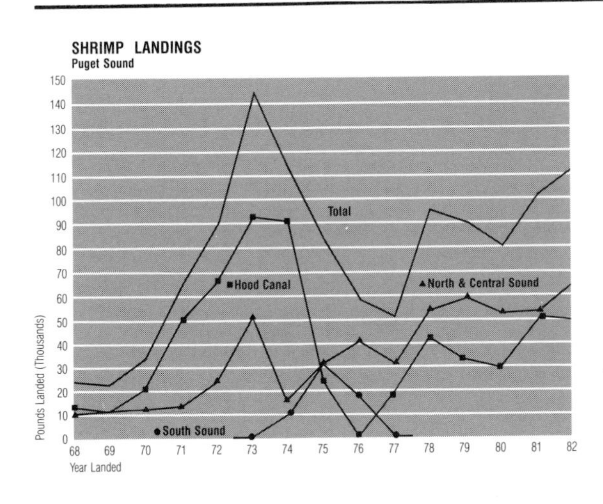


Puget Sound shrimp landings (whole weights). (Source: Annual landings for 1968–82, Washington Department of Fisheries)

Distribution of shrimp harvests in Puget sound, 1976–77. Each circle depicts average annual landings from the value given to the next highest value.

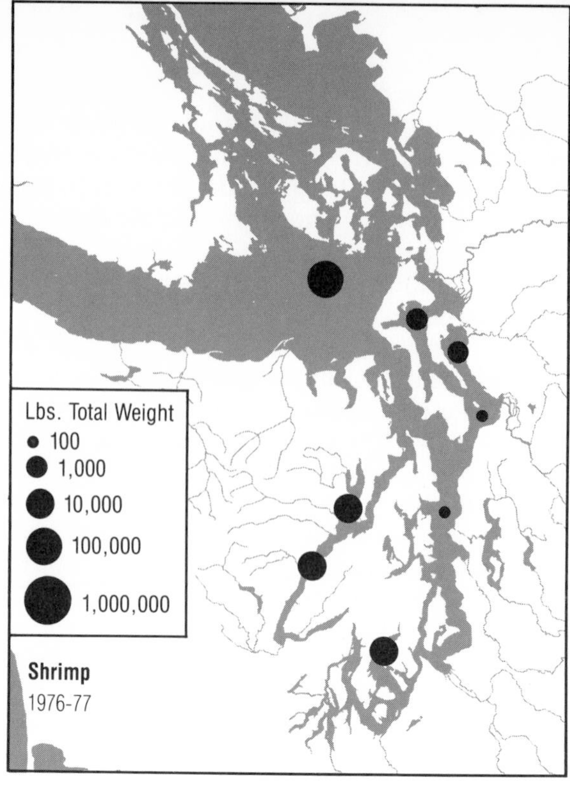


supply of large pandalid shrimp, most of the prawns sold in the retail markets and restaurants of Puget Sound are penaeids imported from the Gulf Coast, Central and South America, and Asia.

## Resource Management

Management of Puget Sound shrimp stocks has focused on comparing catch levels to population estimates, and closing the fishery when populations are low (indicated by low catches) and during spawning and egg-bearing periods. The Washington Department of Fisheries (WDF) estimates stock abundance before the season opens by setting pots at standardized depths and locations, and calculating a catch per unit of effort (CPUE) to provide a population index. These catches also provide data on frequency of abundance versus shrimp length to allow evaluations of year-to-year variation in spawning or recruitment success. After the shrimp season opens, WDF surveys the fishery to determine effort and reviews fishermen's logs to keep track of catch levels and adjust their estimates of abundance according to the CPUE and overall efforts to ensure against overharvest.

These management tools, as well as a requirement for larger mesh

(⅞ inch) in pots, have been most successfully applied in Hood Canal, an area that has experienced increased fishing effort in recent years.

Because there is still no limit to the number of commercial licenses issued, major conflicts remain between commercial and sport fishermen, who are competing for the same shrimp stocks at the same time. Thus, while the amount of gear and commercial pressure could be controlled by a limited entry system, there still would be unresolved problems of overlap of commercial and sport fishing, particularly in Hood Canal. The Department of Fisheries has not as yet been given the authority by the Legislature to develop new management programs that could lead to a solution to these conflicts.

## Expanding the Fishery—Possibilities and Problems

Management efforts to increase the commercial shrimp catch in Puget Sound are limited by the abundance of natural stocks, which are harvested at the level believed to be the maximum they can sustain. Therefore, the potential of increasing shrimp production through artificial cultivation is being explored by the National Marine Fisheries Service and the University of Washington School of Fisheries. Pandalid shrimp have been successfully cultured to the adult stage, and pilot-scale aquaculture systems have been developed. Some species show poor growth and survival when cultured in higher densities, but the species that appears to be best suited for culturing is the spot shrimp, *Pandalus platyceros*. Spot shrimp culture still must be considered an experimental venture, and its commercial feasibility is yet to be established. Nevertheless, the capability of an aquaculture enterprise to produce large, premium quality pandalid shrimp is incentive for continued work leading to commercial culture.

In addition to Pandalid shrimp, there are a number of other shrimp families in Puget Sound with potential commercial importance. These include a number of small shrimp species, such as the little gray shrimp (*Crago* sp.), which are abundant in shallow intertidal waters and tidepools but are considered to be too small and dispersed to harvest commercially. These small species are prized by Oriental consumers, who consider small, light-bodied, whole shrimp to be a delicacy and are willing to pay high prices for them. There is no information on the abundance of these shrimp in Puget Sound, however, and until their populations are assessed, initiation of even a small-scale commercial fishery is not feasible.

There are two shrimp species that are abundant in Puget Sound and large enough to be considered a commercial resource, the ghost shrimp (*Callianassa californiensis*) and the mud shrimp (*Upogebia pugettensis*). The ghost shrimp has a large, soft body but does not look much like a shrimp; the mud shrimp is large and has a more shrimplike

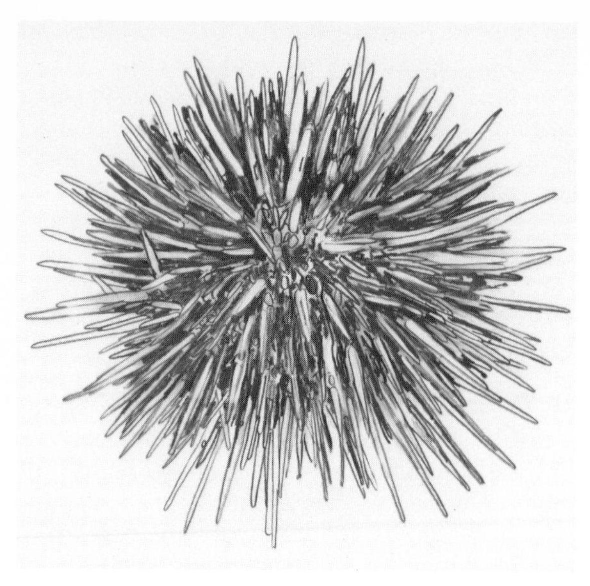

The globular and readily identifiable sea urchin is harvested commercially from northern Puget Sound and the Strait of Juan de Fuca. The red urchin is the largest with a diameter of up to five inches and spines two inches long. Only the gonads, which can account for up to 25 percent of the urchin's weight, are consumed.

appearance. Although oyster growers might welcome commercial harvest of these shrimps, any commercial use is likely to be limited to supplying bait markets, because the shrimp's size and appearance are not well suited to the food market.

## SEA URCHINS

Of the three species of sea urchins common to Puget Sound—the red sea urchin, *Strongylocentrotus franciscanus;* green sea urchin, *S. droebachiensis;* and purple sea urchin, *S. eureureatus*—only the red sea urchin is harvested. It is larger than the other two species and is relatively abundant. The green urchin, while abundant, is usually too small to harvest commercially.

Urchins are adaptive feeders, and though they graze on seaweed where it is plentiful, they can also survive quite well on detrital material and even dissolved organic matter. Where urchins are very abundant (as they are in Southern California), they can overgraze seaweed stocks, which also constitute an economically valuable crop.

Sea urchin populations are affected by food availability, predators, and disease. They are preyed upon by crabs, starfish, finfish, seagulls, and sea otters. Mass mortalities of red sea urchins have been reported off southern and central California, resulting in the rapid recolonization of the areas by brown algae. Such mortalities have practical significance because they can reverse the overgrazing of kelp by large sea urchin populations where kelp is an important crop, or mortalities may threaten production where a new fishery is developing.

## Resource Management and Harvesting

Because sea urchins are an important source of food for many benthic animals, the Washington Department of Fisheries has established restrictions on fishing seasons, areas, size, fishing methods, and species. The regulations are designed to be flexible in ensuring the maintenance of high quality stocks and in meeting the needs of the industry. Management decisions regarding the location and size of the harvest area are based on surveys made before and after the season, and on urchin catch logs and landing tickets filled out by fishermen. Fishing is restricted each year to one of three sectors along the Washington coastline of the Strait of Juan de Fuca from Cape Flattery (Tatoosh Island is closed) to Port Townsend. Harvesting is rotated within permitted fishing areas; open one year, then closed for two years to allow natural recovery of stocks. In addition, harvest has been allowed in the San Juan Islands, except for San Juan Channel and a portion of the west coast of San Juan Island, which are designated as scientific reserves and are closed to fishing to protect sea urchins and other scientifically important species.

Harvesting is restricted to red urchins 3¾ to 5½ inches wide, and green urchins (no size limit); harvest of purple urchins is prohibited. Harvesting is allowed only October through April.

Sea urchins are fished locally by divers using scuba or surface-supplied air in waters beyond 10 feet below mean lower low water. To harvest them divers must use hand-operated equipment that does not penetrate the test (exoskeleton). This reduces harvest efficiency greatly in deeper waters and precludes economical harvest beyond 50 feet. An average daily catch is about 2,000 pounds, although some divers have reported landings as high as 5,000 pounds per day. The size of the fish-

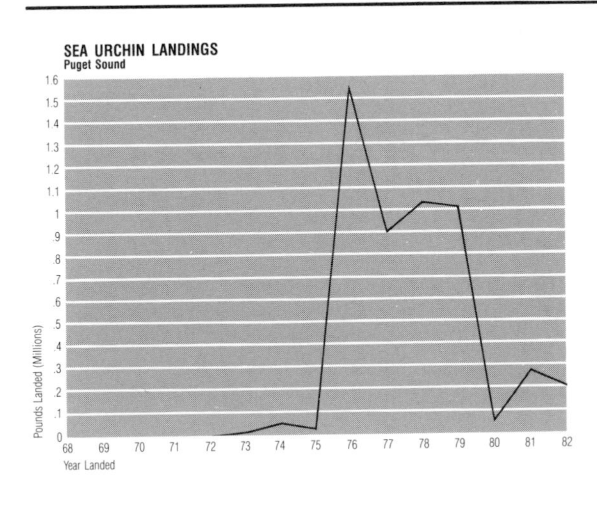


Puget Sound sea urchin landings (whole weights): since 1982, landings have been 200,000–600,000 pounds annually. (Source: Annual landings for 1968–82, Washington Department of Fisheries)

ery is small, with no more than six to eight full-time divers, and two processors (one in Washington and one in Canada).

## Processing and Marketing

The only portion of this animal that is an important commodity is its gonads. Japan presently is the principal world market for sea urchin gonads, called *uni*. There is very little local demand, with the exception of some interest in the local sushi market. The Japanese consume about 4,000 tons of uni annually, much of which is imported. Red sea urchins, *akauni*, are the preferred variety. They are usually served in sushi restaurants, raw, pickled, or mixed with other seafood on top of small vinegar-flavored rice balls.

The sea urchin trade is highly specialized and fraught with technical problems arising from the product's extreme delicacy and quality requirements. The industry is labor intensive because there is no mechanized method for removing and packaging the delicate gonads. They are processed by hand, which involves breaking the shell, extracting and cleaning the gonads, and packing them either fresh, salted, fro-

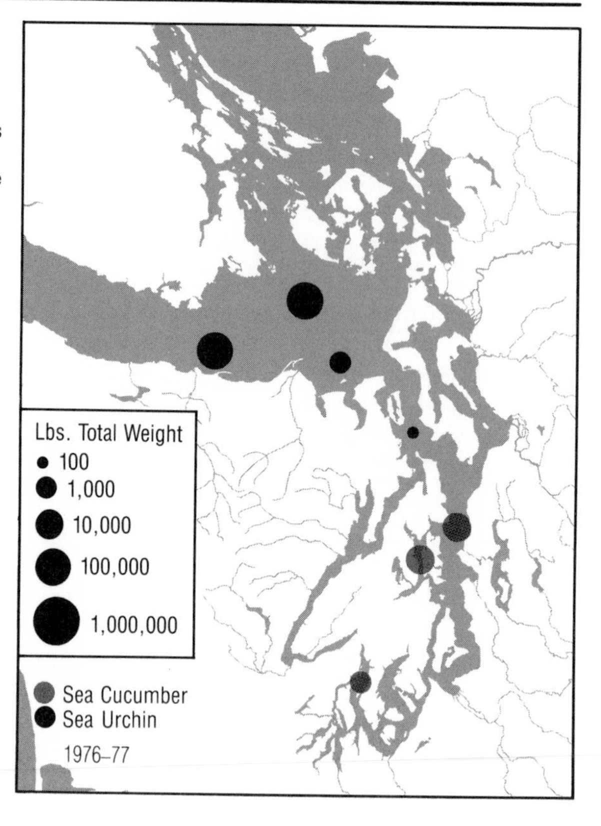


Distribution of sea urchin and sea cucumber harvests in Puget Sound, 1976–77. Each circle depicts average annual landings from the value given to the next highest value.

zen, or canned. They are shipped by air to Japan, usually within a day of harvesting. Finished product yields are low, five percent of the total pounds landed. To ensure the highest price they must be carefully processed to be in top condition and have good color, yellow or orange (dark brown gonads have a greatly reduced value).

Market prices tend to fluctuate greatly. California is the principal domestic competitor and when landings are high from that state (i.e., when the sea is calm), the prices for Washington urchins are depressed. Much of the price fluctuation is due to the vagaries of the Japanese market, which was rather undemanding in the past but in recent years has accepted only urchin roe of the highest quality and freshness. Washington harvested urchin roe commands the highest price in Japan during the Christmas season—a period when Japanese supplies are limited or of poorer quality.

Prior to 1976 only small quantities of sea urchins were being harvested from Puget Sound. In 1971 the catch was only 1,800 pounds; by 1976, it rose to 1.6 million pounds with a value landed of $125,000. Since then the landings have varied widely, and the recent harvests indicate that 500,000 pounds a year is an optimum production level for Washington State (about four percent of the total shellfish landings). The worldwide production of urchins is dominated by Japan (26,000 tons whole weight in 1981), Chile, Korea, and the United States. Sea urchin harvesting will probably continue to depend on the Japanese market. Although demand for Puget Sound sea urchins is high in some years, it is not unlimited and there is intense competition from foreign suppliers.

## Prospects for Expansion

Harvesting and processing technology is unlikely to undergo any radical change. Hand harvesting is inefficient, but it minimizes environmental disturbance and restricts the fishery to shallow waters, thereby retaining a deeper resident population to resupply the harvested areas. Sea urchins have not been cultured commercially, even though they are commonly grown in laboratory aquaria as research animals used for the study of cell development. Mass culturing for commercial production, however, presents economic obstacles associated with food and holding facilities that are insignificant to small-scale laboratory-use production. The only real commercial-level experimentation has been by Japanese scientists who found that the animals grow more than twice as fast using hanging culture techniques similar to those used for bivalves. They also noted that seed can be gathered using suspended collectors, the same technique used to collect oyster and scallop spat.

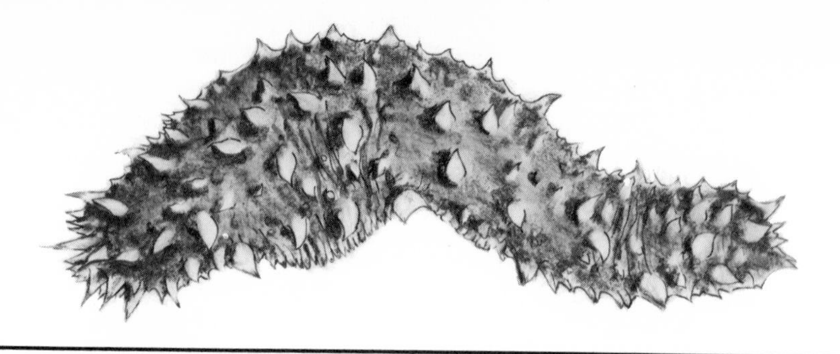

*Stichopus californicus* is a large, common intertidal and subtidal sea cucumber in Puget Sound. It often attains a length of 15 inches, has a reddish brown coloration, and is covered with pointed, fleshy warts. Only the body wall muscles are generally considered to be edible.

# SEA CUCUMBERS

Sea cucumbers are tubular, sedentary or slow-moving creatures that are edible when properly prepared. Of the numerous species found in Puget Sound, the one with the most commercial potential due to its size and abundance is *Stichopus californicus*. At present, the sea cucumber fishery in Puget Sound provides only supplementary income to fishermen. *Stichopus* is generally abundant throughout much of Puget Sound waters and there has been no effort to control the harvest or determine the fisheries potential of local stocks; however, some basic biological information that may assist in management of the fishery is available. The fishery is closed only in the scientific reserve waters surrounding San Juan Island and is monitored by the Washington Department of Fisheries through catch logs and landing tickets filled out by fishermen.

## Harvesting, Processing, and Marketing

The fishing technique for sea cucumbers is relatively simple. They are either hand-harvested by divers or netted in trawls. Sea cucumber production has been a small fraction of total annual shellfish production of Puget Sound (0.5 to 4 percent), and until recently production levels have not changed much in the past decades. They have ranged from 12,600 pounds in 1935 up to about 421,000 pounds in 1980, and 27,000 pounds in 1982.

Whole sea cucumbers are sold either fresh or dried; the intestines are sold sliced, pickled, and dried; and the longitudinal muscles from the body walls are sold prepared and sectioned. Most of these products have little appeal for the average American palate, with the exception of the prepared muscles, which are very tender and taste much like clam meat. Unfortunately, the muscles comprise only a small portion of the whole animal.

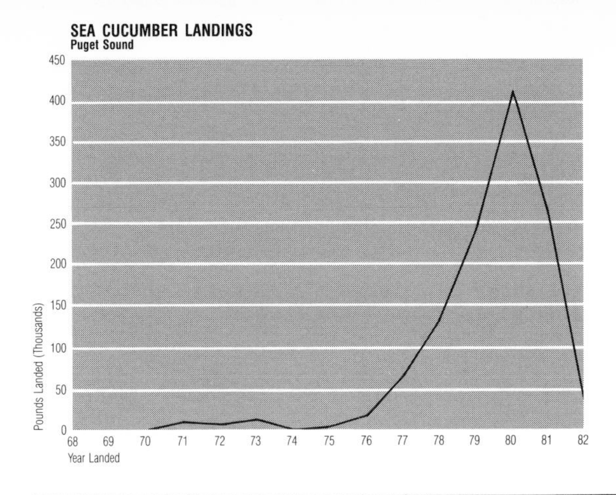


Puget Sound sea cucumber landings (whole weights). (Source: Annual landings for 1968–82, Washington Department of Fisheries)

The most common product made from sea cucumbers is dried cucumber, called *trepang* or *beche de mer*. Dried cucumbers are soaked in water, sliced, and added to soups or stirfry dishes. They have an inoffensive, gelatinous texture and a very mild flavor. Konowata, which consists of a thick, muddy brown mixture containing long, slimy strands of intestine, has less appeal.

Sea cucumbers are widely marketed in Asia, with the greatest demand in China. The most active trading occurs in Singapore and Hong Kong, where cucumbers are imported from other parts of Asia, Africa, and the South Pacific. The market is even more unstable and fickle than that for sea urchin roe.

## Fishery Limitations

Like sea urchins, cucumbers suffer from limited domestic demand. At most, they provide only supplementary income for fishermen and are limited to foreign markets that are already fiercely competitive. In addition, because the animals are of minor commercial importance, there have been few studies aimed at determining their fishery potential. Critical ecological data and information on reproduction, recruitment, growth rates, uptake of pollutants or human pathogens, and interactions with other animals found in the same environment are lacking or very limited for potential commercial species. Additional studies would be needed to initiate and wisely manage this fishery if it were to intensify.

# OTHER SHELLFISH SPECIES

There are several other shellfish species in Puget Sound that have potential commercial value, but few that are suitable for domestic use. Dried jellyfish are popular in China and Southeast Asia (over 22,000 tons were harvested in 1981), and tunicates or sea squirts, have a small market in Korea and Japan and are cultured on rafts in northern Japan. The suitability and availability of local stocks of these shellfish for commercial harvest and export is unknown. Also, because they are not classified as "commercial" species, state research and development funding to encourage a commercial fishery for the animals is not available.

There may be some potential for a marine bait worm industry in the Sound. This industry is presently centered in Maine, which supplies over 90 percent of the saltwater bait worms sold to recreational users in the United States. It is Maine's fourth most valuable fishery, with 1981 production of over one million pounds of sand and blood worms. The fishery is incredibly tenuous, however, because the market is unstable and the worms are fragile. Handling and marketing procedures developed in New England could probably be applied to a Puget Sound fishery. Sand worms are common in certain parts of Puget Sound, but little is known about the extent of local marine worm populations, variations in size and annual crops, growth rates, or sustainable harvest rates.

Many shellfish species have potential application in the ornamental fish trade and as a source of chemicals for processed foods and pharmaceuticals. Examples are sea anemones, sponges, tube worms, and hermit crabs. The harvest and economic potential of these marine species in Puget Sound remains to be explored.

CHAPTER SEVEN

# Seaweeds

Seaweeds are large marine plants that belong to three major divisions: green algae (*Chlorophyta*), brown algae (*Phaeophyta*), and red algae (*Rhodophyta*). They vary in form from filamentous, simple, and branched blades, to very complex structures; and in size from barely visible to the naked eye to as long as 100 feet. They grow primarily in intertidal and subtidal habitats to depths of about 30 feet, attached to cobbles, bedrock, other seaweeds, and even animals. A few are free floating. They absorb nutrients from the water directly through their cell walls rather than from the substrate by way of root systems, like land plants. They require sunlight for photosynthesis, which is controlled by depth, turbidity, cloud cover, and shading.

Over 500 species of seaweeds are found in Washington's marine waters—in fact there are more kinds of kelp (a particular kind of seaweed) than are found anywhere else on earth. (A list of commercially useful seaweeds, and their habitat, distribution, and uses is given in Appendix C.) They play two important roles in the ecosystem. First, as producers of organic carbon, seaweeds are as important as phytoplankton as primary producers in the coastal zone. They are food for snails, amphipods, urchins, and other herbivores, but more importantly, algal material enters the food chain as detritus, broken down bits of organic matter. The second major contribution of marine algae is as habitat. Algal beds provide places where animals can hide from predators, feed, and breed, and form substrate on which marine organisms attach. Although the effects of harvesting large quantities of algae are unknown, they would no doubt be far-ranging, disrupting food supplies and destroying important habitat.

## Seaweed Biology

Reproduction and life histories of algae are complex, and many details are still poorly understood, but what is certain is that there is considerable variation. Seaweeds reproduce either by fragments or by mobile or immobile microscopic spores. Some spores are asexual: they simply settle down, attach, and germinate. Other spores act as sexual gametes: they fuse with another gamete before germinating. Some spores germinate and grow into a plant resembling the parent plant, for

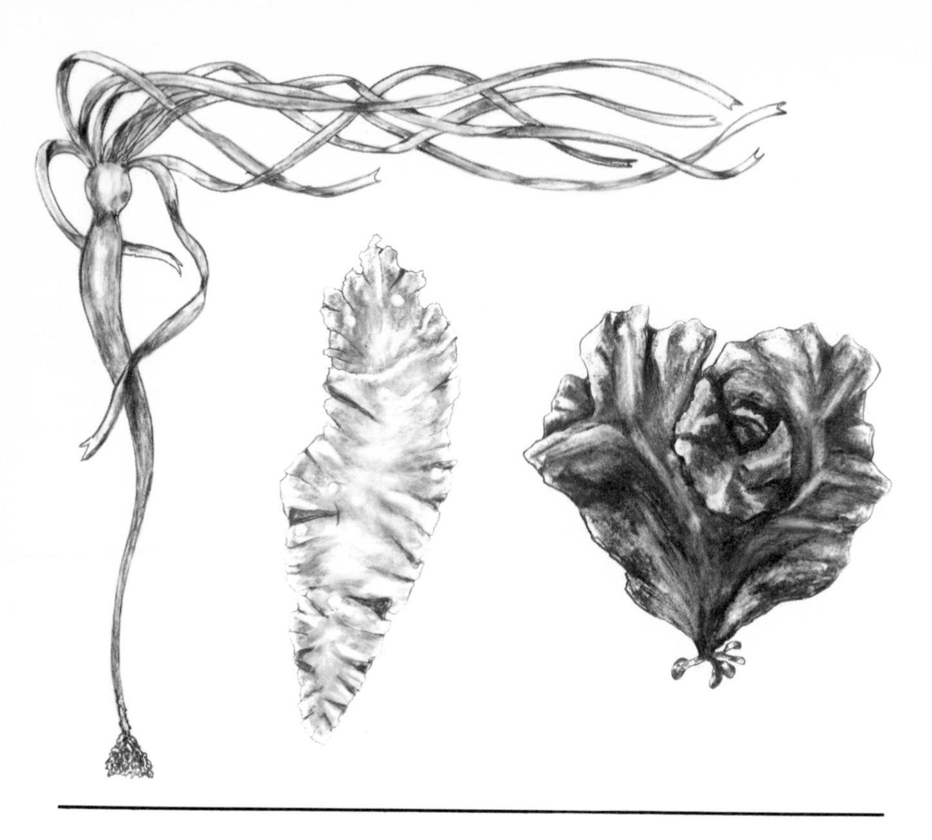

Common Washington seaweeds. Left: *Nereocystis leutkeana* is an important primary producer that can grow over 100 feet long. It has a holdfast, hollow stipe, and blades that float at the water's surface. Middle: *Porphyra abbottae* grows up to a foot long attached to rocks in the intertidal zone. Right: *Iridaea cordata* grows to three feet, is purple-brown with an oily irridescence on the surface.

example, *Iridaea* or *Fucus*; others, such as *Nereocystis*, produce young plants that look unlike the parent, then reproduce in the alternate phase to produce a plant that does. Often, these microscopic spores and germlings are the most delicate part of the life cycle and are easily damaged or disrupted by pollution or degradation of water quality.

Washington's inland waters have ideal conditions for algal growth—excellent water quality, relatively high salinities, and high nutrient levels. Only lack of substrate appropriate for attachment and low levels of ambient sunlight during winter restrict native stock distribution and seasonality. Except for the extreme southern end of Puget Sound, there is generally no nutrient depletion in the summer, unlike surface coastal waters where nutrients may be depleted by rapid phytoplankton growth.

The Strait of Juan de Fuca and the San Juan Archipelago have abundant seaweed stocks in water up to 30 feet with bedrock or large cobble substrate stable enough to be unmoved by currents or waves. Most of the algal biomass is brown algae of the order Laminariales (kelp), especially *Laminaria, Nereocystis* and *Macrocystis*. This biomass is seasonal; the peak algal standing crop occurs in the summer and fall, and there is little or none in the winter.

Areas of Puget Sound with no stable substrate upon which germlings can attach themselves are generally devoid of large seaweeds. Siltation, pollution, and herbivores can also inhibit growth and germling survival. In the San Juan Islands, for example, many areas with rocky substrates suitable for seaweed growth, are instead paved only with crust-forming seaweeds (such as pink coralline algae) that are resistant to grazing by herbivores such as sea urchins and abalone.

## Uses

As part of the marine environment, man the explorer, gatherer, manipulator, and exploiter has always found uses for seaweeds. Coastal-dwelling Native Americans used seaweeds as food, sources of medicinals, and materials for creating tools and containers. For example, the Makah Indians of the Olympic Peninsula smoked and oiled the stipes of *Nereocystis*, then braided them together to make harpoon lines for whaling. *Nereocystis* bulbs were used to store eulachon oil, and steam wood for bending into hooks. Seaweeds were pressed on sunburned lips to cool them, draped over new mothers' breasts to induce lactation, and burned to make acrid smoke that was believed would drive away the harsh winds threatening offshore fishermen.

Algal uses fall into six broad categories: food, fodder and fertilizer, chemicals, pharmaceuticals, phycocolloids, and energy. Some of these uses are minor or outdated, but many others produce little known but vital chemicals critical to modern industry.

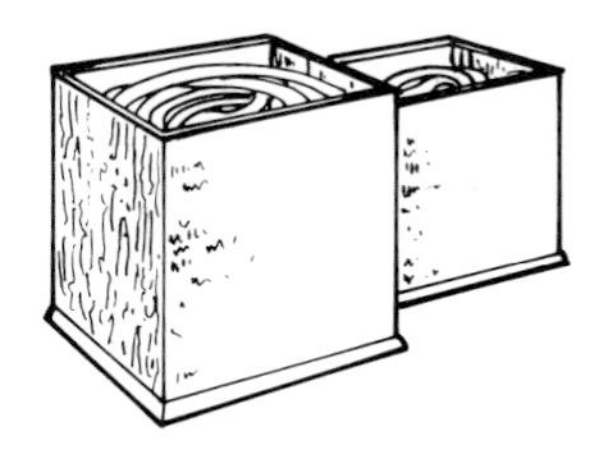

Pacific Northwest coastal Indians used bull kelp (*Nereocystis*) stipes to store eulachon oil. (After Stewart, 1977.)

**Food**

In Indonesia and the Orient—particularly Japan—and among the native peoples of the Pacific Northwest, seaweeds have been gathered for centuries for food and as condiments. As a food resource, seaweeds are nutritionally rich. They contain vitamins A, $B_{12}$, C, D, and E, pantothenic acid, folic and folenic acids, minerals, trace elements, iodine, and protein—as much as 40 percent of total dry weight.

**Fodder and Fertilizer**

Because they are rich in vitamins and minerals, seaweeds have also been used as animal fodder and fertilizer. In Europe and the British Isles, sheep once grazed on beaches, or farmers gathered seaweed to feed to their livestock, attributing to it weight gains, increased egg or milk production, and enhanced resistance to disease. Dry seaweed meal is still used as animal feed supplement.

Liquid seaweed extracts are used as micronutrient-rich soil additives and plant foods that enhance resistance to insects and fungi, increase nutrient uptake, and provide a variety of hormone-mediated responses such as decrease in fruit drop.

**Chemicals**

In the seventeenth and early eighteenth centuries large brown seaweeds were burned to recover soda (sodium carbonate) from the ash, which is used to manufacture glass. The word "kelp" originally meant the ash from burned seaweed, and only later came to mean the seaweed itself. Potash, another product of kelp, was valued as a fertilizer and for industrial processes, such as the manufacture of soap and explosives. Early in the twentieth century potash production was effectively monopolized by Germany, and as political problems in Europe intensified, the United States was forced to develop domestic sources. Kelp resources were inventoried from Baja California to the Aleutian Islands, and large-scale harvesters were built. During World War I, annual production of potash reached 4,000–5,000 tons. After the war, production of alginates began and these products had sufficient value that other uses were no longer profitable. Kelp was a major source of iodine in the 1800s before Chilean guano deposits provided new sources.

**Pharmaceuticals**

As medicine, seaweeds were used to treat goiter and worms in humans. Today, seaweeds and their extracts are used in many applications:

- Calcium alginate is used to dress wounds because it forms a gel and retards bleeding.
- Carrageenan and other red algae extracts (*Constantinea* and *Cryptosiphonia*) have antiviral properties.
- Carrageenan is used to treat ulcers, and to lower blood cholesterol

and serum lipid levels.

- Alginates have been used to treat poisonings due to radioactive strontium and heavy metals, such as lead.
- Dried *Laminaria* stipes swell when moistened, and are used to dilate the cervix before abortion and to widen fistulae and wound openings.
- Carrageenan and agar are used as binder in tablets.
- Agar is used as a base for suppositories, pills, plasters, and ointments, and as an agent for growing bacteria for medical laboratories. Recent shortages have raised the price so that recycling agar is now cost-effective.

**Phycocolloids**

By far the largest use of seaweeds in the western world is the extraction of algal polysaccharides or phycocolloids. Phycocolloids are sulfated polymers chemically related to cellulose. In water they form gels used to give foods such as puddings, ice cream, and sauces a smooth consistency, to prevent ice crystals from forming, and to keep the components from separating.

Phycocolloids also have many industrial uses. They prevent paints from separating and enable them to flow smoothly, yet not drip, and are used in gels for high-speed printing inks. An estimated $200 million worth of algal polysaccharides are used in the United States annually to produce $22 billion worth of goods. There are three major groups of phycocolloids:

*Porphyra* is prepared for eating by processing the wet seaweed in a paper-making machine into these 7½ × 8¼-inch sheets. Over 12 billion of these sheets were produced in Japan in the 1984–85 growing season, with a wholesale value of about $600 million. This bundle of 100 folded sheets is worth about $5.00 to the farmer and $50 to the retail merchant.

- Agar, extracted from various red seaweeds, forms gels and is used in foods and bacteriological plate preparations in medical laboratories;
- Carrageenan, extracted from red seaweeds, forms gels or highly viscous solutions for foods, especially milk products;
- Algin (or alginic acid) from brown seaweeds forms gels or viscous solutions used in industrial and food processes.

These compounds' physical properties vary widely; each species has different types of polysaccharides.

**Energy**

The search for renewable and relatively nonpolluting energy resources has lead to research into large-scale production and conversion of plant biomass to methane gas. The American Gas Association, Energy Resources and Development Administration, and General Electric Company evaluated the feasibility of growing the giant kelp *Macrocystis* for organic matter that could be decomposed anerobically to produce methane. Biological, engineering, and economic problems are inherent to such large-scale production, however. It may require more energy to build and maintain large rope structures (up to 400 square miles) and pump water to provide plant nutrients than could be produced. Nevertheless, *Macrocystis* produces more organic matter per area than almost any other plant, growing in excess of two feet per day under optimal conditions. Research is now focused on creating artificial kelp beds in nearshore waters.

Puget Sound is not an ideal place for biomass energy production, however. First, there is not enough room: only a small percentage of Washington's energy needs could be provided for, even if the entire Sound were utilized for the growing area. Also, the distinct seasonality of growth is a major problem. At the latitude of Puget Sound, only one-seventh of the sunshine in the summer is received in the winter. Although algal growth rates are very high during the summer months, growth almost stops in the winter, when energy needs are greater. As energy costs continue to rise, kelp as an energy source may someday become a reality.

## Harvesting of Wild Stocks

Seaweeds are harvested by a variety of methods, some simple, some sophisticated. In Canada, to collect Irish moss (*Chondrus crispus*, used for carrageenan extraction), loose seaweeds are collected by men standing chest deep in the surf wielding dip nets on long poles, or using baskets dragged through the surf with horses or tractors. Subtidal beds are harvested with toothed trawls dragged behind fishing boats or men working long-handled rakes from small boats in shallow water. In

Japan, konbu (*Laminaria*) is harvested with long, weighted poles with toggles that are twisted to tangle the kelp. *Gelidium*, used for agar, is gathered in Japan and California by divers. Dulse (*Palmaria palmata*) is collected from rocks by hand at low tide.

The alginate industry uses large-scale harvesting methods. In southern California, offshore beds of *Macrocystis* are harvested by ships fitted with underwater cutter bars that cut the kelp about three feet below the surface and a conveyor belt that moves it into the hold of the ship. Smaller barges, fitted with similar cutters, harvest *Ascophyllum* (rockweed) in Nova Scotia, Canada, and Iceland.

During the 1920s and 1930s, several attempts were made to commercially harvest kelp for algin extraction in the San Juan Islands. After an abandoned pea cannery at Friday Harbor was purchased for the extraction plant, however, it was discovered that the *Nereocystis* plant did not regenerate after harvesting as did *Macrocystis* plants in California. The project was abandoned, but through the years, small amounts of kelp have been harvested for making kelp pickles, soap, and fertilizers.

In Puget Sound, the Lummi Indian Business Council experimented with harvesting natural beds of the carrageenophyte *Iridaea cordata* in 1969. Waters near Bellingham and the San Juan Islands were surveyed and the standing crop estimated at 350 to 500 tons dry weight. Divers were trained, funding arranged, and equipment assembled for a full-scale harvest during the late spring and early summer of 1970. The attempt proved unsuccessful, however, because subtidal beds were widely scattered and harvesting was labor-intensive and expensive. Only 30 tons (wet weight) of *Iridaea* were harvested and operations were suspended at the end of the summer.

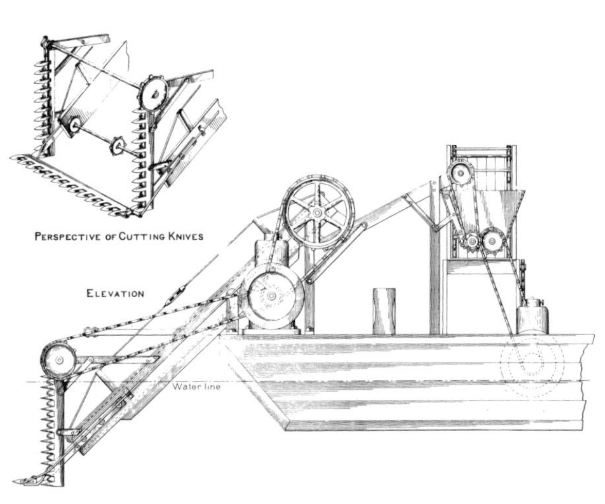


A kelp harvester cutter mechanism mounted on the front of a barge used to cut floating kelp such as *Nereocystis* and *Macrocystis*. This machine was used in the early 1900s, but similar cutters are still used in California today. (Source: Cameron, 1915).

As of 1985, there was only one approved commercial harvesting operation from state-owned lands, Olympic Sea Gardens of Sequim, which harvests edible seaweeds from the Miller Peninsula and dries them and packages them for the health food market. An increasingly large amount of seaweed is being harvested in Washington for food, mostly by Southeast Asian refugees working along the east side of Puget Sound, particularly near Seattle and Tacoma. Although it is uncertain how much has been harvested, in certain areas, such as Titlow Beach near Tacoma, such harvesting has clearly impacted standing stocks.

## The Development of Seaweed Culture

The Japanese are the most sophisticated cultivators of seaweed. They originally gathered species like nori (*Porphyra* sp.), wakame (*Undaria*), and konbu (*Laminaria*) from wild stocks, and only gradually began to cultivate them to enhance natural beds. Early harvesters discovered that many desirable seaweeds such as konbu grow best on "new" surfaces, so they dynamited rocks to provide such surfaces. The next step was to put out new artificial substrates such as cement blocks. Then it was reasoned: Why not help the seeding along by attaching fertile fronds to the blocks before placing them in the ocean? Subsequently, less cumbersome, less expensive substrates were developed, particularly ropes and nets attached to buoys.

### Net and Longline Culture

Today, nori, most wakame and some konbu are produced on nets and longlines seeded with spores of algae selected for good taste and fast growth and placed in bays or the open ocean to grow. Yields are enhanced by restricting when, where, and how much seaweed is harvested, and by weeding and eliminating herbivores, such as urchins, abalones, and snails.

The production of nori is a good example of a gradual evolution of culture techniques. It was once picked from rocks and submerged brush by hand, but by the beginning of the seventeenth century, farmers began to stick brush in shallow bays where they became colonized with attached nori. This gradually led to more sophisticated collection apparatus, and today the most common method is to place artificially seeded, horizontally strung nets between poles or from floating rafts.

Until recently, a major obstacle to true nori culture remained—the source of the spores that seeded the nets was unknown. While it was known that in certain bays during the months of October and November nets could be placed for seeding, these sites were limited. A breakthrough came in 1949, when Kathleen Drew, a British biologist, discovered a probable source of the spores. A small filamentous plant, then

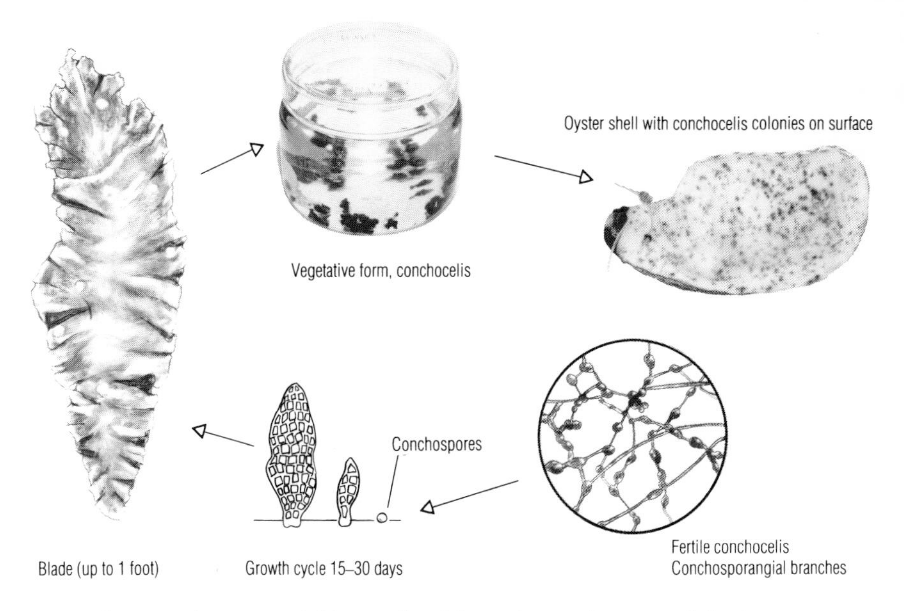


Large-scale commercial farms of nori depend upon the successful culture of two distinct phases in *Porphyra* life history—the conchocelis phase, grown during the summer; and the blade phase, grown during the winter, which is the harvested phase.

called *Conchocelis*, was shown to be produced from spores from the blades. The *Conchocelis* plant was actually a life-history phase of *Porphyra*.

Within a few years the Japanese were growing *Conchocelis* in large quantities, and were producing spores to seed nets when and where they wanted. This discovery emphasized the importance of understanding the biology of cultured seaweeds. The Japanese continue to investigate strain selection, genetics, effects of pollution, fertilization, and effects of growth hormones to manipulate the species and enhance culture.

**Tank Culture**

Culture techniques tested in the United States and Canada show that loose seaweeds placed in tanks of agitated seawater also grow well. The tanks are constructed with cement or are plastic-lined structures on land, or bags hung from floating rafts. Water is agitated by large paddle wheels or compressed air pumped into the bottom of the tank or bag. Some seaweeds, such as Irish moss, fragment in the tanks and self-perpetuate, while others must have fresh inoculum (young plants) provided. To harvest from tanks, excess growth is skimmed off. The advantages of using this method are that growth rates equal or exceed those from natural stocks and pure cultures can be grown. The disadvantages are contamination by weeds, high construction and operating costs for

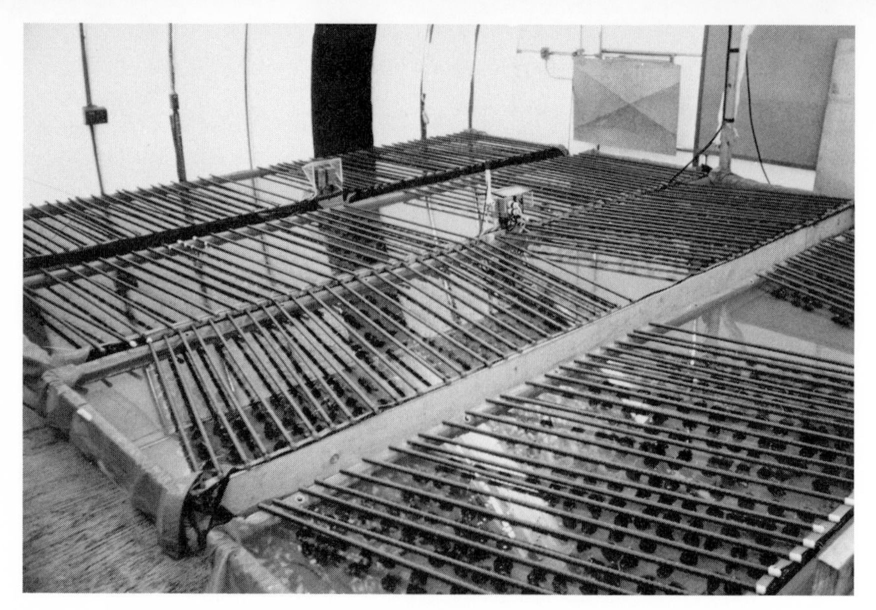

Oyster shells inoculated with conchocelis are hung in pairs in tanks. Seawater in the tanks is temperature controlled, and light levels in the shed are lowered by overhead curtains. The conchocelis grows from April until October.

Newly seeded nori nets are placed 5–10 deep in these floating nursery frames at Squaxin Island. The nets are raised periodically to be dried and to kill unwanted plants and animals. They are left in these frames for about 40 days until the plants reach a third to three-quarters of an inch in length.

Close up of *Porphyra* blades nearly ready for harvesting.

water and air pumps, and the large area of valuable nearshore land needed.

**Seaweed Culture in Washington**

In 1970 the Washington Department of Natural Resources initiated research to develop methods for culturing *Iridaea* on artificial substrates. The first attempts to seed and plant ropes were unsuccessful, but in 1974 ropes and nets were placed over a natural bed of *Iridaea*, and large standing crops were produced. In 1976, the work was expanded to a pilot-scale project. It became evident, however, that natural seeding was unpredictable so seeding was attempted under controlled conditions in tanks with both *Iridaea* and *Gigartina*. The new seeding technique worked, but high production costs caused DNR to drop these species from the project. Nori proved promising for aquaculture development, however, so a Japanese nori aquaculurist, Makoto Inayoshi, was hired to teach DNR project personnel how to grow it. Pilot-scale seeding of nets was accomplished in September 1982 and by October the project had its first successful harvest. Small amounts of nori were frozen and flown to Japan to be processed into sheets, which turned out to be of moderately good quality. Since 1982, the DNR, project has successfully grown three seasons of the conchocelis phase, seeded nets with nori, and grown them to harvest size.

When the *Porphyra* blades reach a length of a third to three-quarters of an inch they are removed from the nursery frames and placed in these production frames. They remain here for another 30 days until the plants reach a length of six to eight inches and are then harvested.

## Economics of the Seaweed Industry

### Wild Stocks

Before natural beds can be utilized, the extent and location of harvestable stocks must be inventoried. The amount that can be reasonably removed is then estimated, taking into account water depth, exposure to waves and storms, seasonality of growth, standing crop, and yearly variation of biomass. Short- and long-term harvesting effects are assessed for the seaweed itself and associated flora and fauna. The percentage of foreign weed and debris in the harvested material is estimated to determine quality and price. The number and size of vessels required for harvest is determined by site selection, considering distance from the harvesting areas to a shore facility for unloading or to a processing plant.

About 900,000 dry tons of seaweeds were harvested worldwide in 1978 for the world phycocolloid market, which is worth an estimated $1 billion per year. The U.S. carrageenan market alone is worth $68 to $100 million per year. The limit to market growth has been the lack of a stable supply of raw materials. Until the industry has a dependable supply of phycocolloids, it will be reluctant to make a large product line depending on algal gels. The market for phycocolloids is increasing only 10 percent per year; however, the market for farm-produced (cultured) seaweeds may favor an annual increase of 25–30 percent. In the case of carrageenophytes, the market is a "buyer's market"—the number of companies purchasing raw materials is very small.

The value of natural beds of *Macrocystis* and *Nereocystis* in Puget Sound is calculated from standing crop estimated by F.K. Cameron in 1915 for the Department of Agriculture: 353,000 fresh weight metric tons (30,000 dry weight metric tons) of kelp in the Strait of Juan de Fuca and Puget Sound. At a dry weight value of $160 per metric ton (1977), the total harvestable value is $4.8 million. At a 19 percent algin yield worth $2.30 a pound, the total processed value is $28.8 million.

The Lummi Indian Council estimated the standing crop of *Iridaea cordata* to be between 318 and 454 dry weight metric tons in the San Juan Islands, which, at the 1982 market value of $400 per dry metric ton, yielded a value of $127,200 to $181,600. The commercial yield of carrageenan is 35 percent of dry weight after processing with a value of $5.00 per pound; the processed extract was valued between $1.2 and $1.75 million.

### Cultured Seaweeds

Costs can be estimated for Japanese culture methods, but for other culture methods costs are proprietary information so exact figures are unknown and likely to remain so. As with the harvesting of natural beds, the culture site is critical: proximity to a processing plant or loading facility is desirable to curb transportation costs, and on-site personnel may be needed to prevent vandalism and to repair damage from storms and other natural phenomena. Tank-agitated culture is still in the developmental stages, therefore costs are high. As the technology improves, costs should decline.

The Japanese grow an estimated 2.1 million tons of seaweeds annually (wet weight, 1971–73 harvest averages); about half is from native stock, the other half is cultivated. This amount is only 12 percent of the estimated 17.2 million tons of harvestable algae potentially available.

The total amount of edible seaweeds produced yearly is difficult to determine because much of it goes through small foreign markets. Japan produced an estimated 654,000 dry metric tons of nori, wakame, and konbu, worth $563 million in 1973. In 1980–81, 24,900 dry metric tons of nori were produced in Japan, with a value of $491 million to the farmers.

## Production Constraints and Incentives

### Standing Stock Limitations

The Washington State Department of Natural Resources, under RCW 79.68.080, has authority to "develop and improve production and

These nori nets in Matsushima Bay, Japan, illustrate a traditional Japanese method of nori production: growing the blades on nets spread horizontally between poles driven into the bottom of shallow bays. As a result of widespread landfill activity and pollution, this is no longer a common scene in Japan.

harvesting of seaweeds," and it has developed regulations for removal of seaweeds from bedlands and state-owned tidelands (WAC 332-30-154). (These regulations and the commercial species established as part of these regulations are given in Appendix D.) Seaweed growing in the water or detached plants washed up on privately owned or leased tidelands in the state of Washington belong to the person owning or leasing those tidelands. Seaweeds growing in the Marine Biological Preserve established in all of San Juan County and Cypress Island, Skagit County, are protected and may be collected only with the permission of the director of the Friday Harbor Laboratories. Exceptions are any organisms "gathered for human food and the plant *Nereocystis*, commonly called kelp."

Cameron's 1915 estimate of 353,000 wet metric tons of *Macrocystis* and *Nereocystis* in the Puget Sound region is approximately ten times what is needed to establish a processing plant. This standing crop was never verified using modern techniques, nor was harvest shown to be truly profitable considering harvesting techniques, stock dispersal, and difficulties in harvesting some sites. Royalties to the state to harvest this resource would have to pay for inventory and management. The effects of harvests on the kelp and the kelp bed's role as a food source and habitat are not known, and the costs of making this determination would also enter into the true cost of harvesting.

**Institutional Constraints**

The National Research Council has observed that "constraints on the orderly development of aquaculture tend to be political and administrative rather than scientific and technical." Indeed, although the biological feasibility of growing *Porphyra* on nets, *Laminaria* on longlines, and *Gracilaria* and *Gelidium* in tanks has been shown, there are no commercial ventures growing these seaweeds in the Northwest. The constraints seem to fall into two broad categories—permits and marketing.

The permitting procedure in Washington for aquaculture projects is usually long, costly, and very uncertain as to its outcome. A shoreline permit is first required from county government, and it is at this level that most opposition is heard. While opponents may state their objections in environmental terms, their real argument appears to be aesthetic. Many people approve of the idea of aquaculture, but those who own property near the water decry any change in "their" marine view. While appeals to the State Shorelines Hearings Board have not supported aesthetic arguments, the delay and costs of legal counsel are costly to a project.

A permit is also required from the U.S. Army Corps of Engineers in order to place any structure in navigable waters. Drawn plans must be circulated to state and federal agencies for review, and comments gathered and reviewed by the Corps. Environmental concerns are usually resolved in the course of this process, and a project may be modified or mitigated to prevent environmental damage.

Further requirements are a hydraulics permit from the Washington Department of Fisheries, which assures that the project will not damage fish habitat or stocks, and a lease from the Department of Natural Resources if the proposed project is on state-owned tidelands or bedlands. Rates for the lease are negotiable with the department.

**Finances and Marketing**

Because of the length of time that may be required to obtain permits, and because there is no way to tell how long the permitting process will take or even whether permits will be issued at all, investors have not been eager to support aquaculture projects. Financial backers are also hesitant about such projects because management skills are often lacking in people who want to start an aquaculture business. Successful aquaculture ventures consist of a team of skilled business managers, technicians, and accountants—or a rare individual with all these skills.

At the Washington Department of Natural Resource's nori aquaculture research site near McNeil Island, *Porphyra* is harvested from the nets by a cutter mounted in a boat. The boat is run under the nets and the seaweed is cut off and drops into baskets under the cutter. (Photo by Jennifer Werner, courtesy of the Seattle Post-Intelligencer)

Determining the market for a product is often difficult and expensive, yet marketing is the first thing that a potential aquaculturist must investigate. A marketing strategy must be devised and carried out. While these problems are obvious, many technically oriented aquaculturists may wait until well into a project before dealing with these aspects of the business.

In 1985, the Washington State Legislature passed legislation that places the responsibility for marketing of aquaculture products with the Washington Department of Agriculture (ESB 3067). In the past, no agency was charged with this responsibility. The bill treats aquaculture products as agriculture products, and allows the formation of a marketing commission for aquaculture products.

# Epilogue

This account of shellfish and seaweed harvests in Puget Sound began with a brief look at the past—the long history of harvests predating European and American colonization of the region by hundreds of years. Present production of shellfish and seaweed is dominated by harvests of bivalve molluscs, including aquaculture of oysters, clams, and mussels, and capture fisheries for natural clam and geoduck stocks. Crab and shrimp fisheries constitute a smaller but important fraction of total production. Marine species such as squid, octopus, sea urchins, and sea cucumbers harvested from Puget Sound at low levels expanded their market in the early 1980s; but played only a minor role in total harvests. Finally, seaweeds have been harvested in increasing quantities in a personal use fishery and considered for commercial production.

## The Capture Fisheries

Capture fisheries, an extension of the hunting-gathering tradition, are strictly regulated by state management groups. Capture fisheries range from traditional crab-pot fisheries to the diver harvest of geoducks. They also include hand-picking of seafood and bait products, such as seaweeds from intertidal rocks.

Variations in landings and shifts in shellfish species taken in capture fisheries over time reflect rapid development and exploitation, often resulting in overharvesting of stocks, followed by precipitious declines in production when a stock was unable to reproduce at a rate required to sustain the harvest effort. Such situations recurred throughout Puget Sound in the scallop and shrimp fisheries.

Differences in species life histories and growout rates at different locations in Puget Sound present complex management problems, especially when the goal is to ensure uniform production from year to year. Regulation and management that accomplish a stable harvest reflect an understanding of the dynamic nature of fisheries stocks gained from management experience, research, and the knowledge gained from the fishermen themselves.

## Aquaculture

Culture of oysters has a long history in Puget Sound and by the early 1980s, commercial aquaculture techniques had been adapted to rearing of clams and mussels as well. Pilot-scale culture of marine species such as seaweeds, geoducks, abalone, and scallops have also succeeded. Shellfish aquaculture—still dominated by oysters, clams, and mussels—accounts for a large share of Washington State's total shellfish production and is exceeded only by the geoduck harvest in Puget Sound.

## Issues and Development

Capture fisheries are generally managed at the state level with a relatively straightforward regulatory process controlling harvest rates (except when nearshore substrate is likely to be disturbed, as in the case of mechanical clam harvesting). Aquaculture is affected by federal, state, county, and local interests and is subject to a plethora of management controls. Aquaculture development permits are required for construction of shoreside facilities, digging of wells, and control of aquatic cultivation activities affecting the marine environment. As a result of special environmental legislation to ensure protection for sensitive marine environments, far more permits are required for projects near or on marine water bodies than are required for more landward operations.

Shellfish scientists and managers recognize many of the critical biological and environmental factors affecting shellfish production. Similarities in the biology of marine species link many species in the capture fisheries to those that are farmed. Most have free-swimming stages in their life history that intermix in the water column, thereby creating competition for food and space. Many edible and not-so-edible shellfish (clams versus barnacles, for example) filter similar food particles from the water. Infectious diseases and pests occur in plants and animals reared on aquatic farms as well as in species taken in capture fisheries. Many of these pests are native to wild populations, and some of them—such as oyster drills—were imported along with commercial shellfish introduced into state waters.

Capture fisheries and aquatic farms both depend upon good water quality—a dependency that creates economic incentives to preserve and improve water quality. Aquatic farming is also uniquely dependent on proper maintenance and enforcement of regional water quality standards. Problems caused by contamination of shellfish beds all too often can be traced to nonenforcement of existing laws, lax or improper application of land use codes, little or no inspection of onsite facilities, and lack of upland facility maintenance.

Just as there are significant similarities between aquaculture and capture fisheries, there are similarities between aquaculture and agriculture. Aquaculture allows sustained production of uniform, high-quality seafoods that can be tailored to the market. It is a private venture—a cultured oyster is owned by the aquatic farmer throughout its growth, not just at the time of harvest. In many cases aquatic seabeds are also privately held. Management of aquaculture production has been largely left to the skills of the farmer, although state and federal organizations have provided many services to growers. In recognition of the similarities between aquaculture and agriculture, many aquaculture functions (as defined by the State) have been placed under the State Department of Agriculture, and products grown by aquatic farmers are now classified as agricultural commodities.

Probably the most critical management issue remaining in both shellfish and seaweed fisheries and culture is that of user conflicts. These are some typical examples:

- *Conflicts between aquaculture and capture fisheries*—Many sites suitable for seaweed (nori) culture and raft culture of shellfish may be in waters less than 60 feet deep. Geoduck clams also frequently occur in such areas and may be difficult to harvest when overlying waters are used for aquaculture.
- *Conflicts between capture fisheries*—Locations where crab pots or traps are set can coincide with those areas that are used by commercial purse seine and trawl vessels. This has resulted in loss or damage to pots or the incidental capture of crabs in the nets of the fishing vessels.
- *Conflicts with other user groups*—A battle, couched in environmental terms, is looming between competing users of the limited nearshore resources and space. This often places waterfront owners, with little direct economic interest (other than a perceived loss of property values) pitted against those who are attempting to gain a livelihood from the sea. The central issue appears to be aesthetics—people do not want their views interrupted.

## Future of Capture Fisheries and Aquaculture

The striking feature of the shellfish and seaweeds harvested from Puget Sound is not production volume (which is low compared to finfish), but in the diverse nature of the products and the often unusual and special markets that they enjoy. Markets are driving these industries and the demand for shellfish appears to be extremely strong. This is in large part due to the increasing consumption of seafood in Washington State, excellent quality of Puget Sound harvested and grown products, and long-term demand for the products in other west coast areas, particularly California. To meet this demand, more cooperative efforts will be needed, particularly in marketing and purchasing, which will eventually lead to less room for independent competition.

Technically there is great promise for increased aquaculture production, especially in the use of artificially enhanced stocks and more intensive culture methods. In the past few years sophisticated systems have developed to culture algae and mussels; to rear oysters and clams in hatcheries; to manage and maintain water quality in marine culture systems; and to control disease. Future advancements will depend on research and increased availability of funds to support it. Expansion of an infrastructure is required similar to that needed in agriculture. Beyond technological solutions, there is a need to encourage a nurturing attitude, and a willingness to work long hours—aquatic crops cannot be put out and forgotten, or left to grow until the next season.

In an era of heightened environmental awareness, issues that have not been of great concern in the past will become dominant in shaping the industry. With increased human population in Puget Sound and consequent pollutant loading from domestic, farm, and industrial wastes, hard choices will have to be made concerning the tradeoffs between economic and social contributions of capture fisheries and aquaculture and the costs of controlling landside wastes. Harvesters and growers may find their collection techniques questioned, and adverse affects of harvesting and culture will have to be resolved before harvesting can proceed—these answers will be costly and come slowly.

## Outlook for Puget Sound Capture Fisheries

The prospects for Puget Sound's capture fisheres are fair. Cycles of discovery, overexploitation, and stock crash will have to be avoided in order to sustain these fisheries for future generations, however. Major fluctuations in shellfish production is unlikely in the near future because most stocks are fished at maximum sustainable yield and cannot withstand greater fishing effort. Existing stocks are managed to restrict total harvest levels; often, this is accomplished by limiting entry in the fisheries. The inability to increase production of wild stocks of shellfish tends to tip the production sector away from capture fisheries and toward aquaculture.

## Outlook for Aquaculture

The outlook for aquaculture ventures is good, provided they are allowed a favorable economic, environmental, and regulatory climate. Near-term production increases are predicted for oysters, clams, mussels, and marine plants by reusing old culture areas for new crops, applying better crop management methods, and expanding into new culture areas. The expansion of aquaculture will probably be predominantly in economically depressed areas. Employment opportunities in aquaculture could occur in such areas as South Puget Sound, east of Whidbey Island, and Hood Canal.

# Appendices

# Appendix A

## Physical and biological characteristics of Puget Sound shellfish

OYSTERS
Physical and Biological Characteristics

| | |
|---|---|
| **Distribution and Zonation** | |
| Distribution | *Pacific Oysters*—Found throughout Puget Sound. Natural populations occur in Japan, and they have been transplanted to many temperate to tropical waters.<br>*Olympia Oysters*—Native to Puget Sound. Once common, they are now abundant only on a few commercial beds in the South Sound. Also occur in Willapa Bay, California, Alaska, and British Columbia. |
| Zonation | *Pacific Oysters*—Occur commonly from 2 feet below high tide to lower low water, or may be found to -50 feet, on the surface or partially buried in coarse to muddy substrates. Most commercial harvest occurs between 3.5 and -1.5 (feet) tides.<br>*Olympia Oysters*—Found between mean tide and low tide levels; uncommon subtidally. Early natural beds were found in pools at low tide, usually on the surface of mudflats and gravel bars near the mouths of rivers or streams. |
| Substrate | *Pacific Oysters*—Solid rock to mud; prefer to settle on old shell.<br>*Olympia Oysters*—Prefer firm bottoms of pea gravel or coarse silt overlain with fragments of shell; require clean substrate for setting; may be seen singly or clustered on rocks, or pilings. |
| **Physical/Chemical Tolerances** | |
| Temperature | *Pacific Oysters*—Optimum: 50–85°F; will tolerate near-freezing conditions but do not grow.<br>*Olympia Oysters*—Optimum: 56–65°F; will die if exposed to freezing or elevated temperatures. Seasonal extremes may result in poor setting and survival. |
| Salinity | *Pacific Oysters*—Optimum: 10–30 ppt; will tolerate 5–37 ppt, and can be held at 0 ppt for a short period of time. Tolerance decreases with increasing temperature.<br>*Olympia Oysters*—Require slightly more saline conditions for optimal growth than Pacific oysters. |
| Dissolved Oxygen | Can withstand low oxygen levels for several days, and anaerobic conditions for a short period of time. Tolerance decreases with increasing temperature. |
| pH, Ammonia, $H_2S$ | pH not usually a problem—acceptable range is pH 8–9; increased ammonia and $H_2S$ identified as causes of decreased growth and poor condition; however, lethal concentrations of ammonia are high (100–900 ppm). Oysters are usually more sensitive during early development stages. |
| Currents | Tend to grow better in areas with moderate currents. |

**Reproduction**

| | |
|---|---|
| Type of Reproduction | *Pacific Oysters*—Protandric hermaphrodites; sex varies with food supply and age. Older or well-fed individuals are usually female. Reproduce by external fertilization.<br>*Olympia Oysters*—Protandric hermaphrodites; internal fertilization. Function as males in early part of season, then as females; may hold over during the winter in transition between male and female. |
| Natural Spawning Season | *Pacific Oysters*—Late June to early September, usually late July to early August. In the Puget Sound area, natural spawning is infrequent except in Hood Canal.<br>*Olympia Oysters*—Usually in summer but may extend for up to six months from late spring to fall. |
| Environmental Limits | *Pacific Oysters*—Begin to mature in March and spawn when temperature exceeds 65–70°F. Optimum salinity at spawning: 20–25 ppt. Release of sperm is triggered by temperature rise of about 8°F and the oysters resorb eggs and sperm if optimum temperature is not reached during the spawning season. Optimal spawning conditions are when temperatures reach about 68°F for at least three weeks.<br>*Olympia Oysters*—Spawn when water temperatures exceed 56°F. |

**Growth**

| | |
|---|---|
| Type of Reproduction | *Pacific Oysters*—Pelagic eggs; planktonic larvae. Juveniles permanently attach to adult shell or bottom material. Sexually mature by first year.<br>*Olympia Oysters*—Eggs develop and hatch internally, planktonic larvae, permanently attached from juvenile to adult, sexually mature by the second year. |
| Time for Egg Development | *Pacific Oysters*—About 48 hours.<br>*Olympia Oysters*—9–11 days. |
| Time for Larval Development | *Pacific Oysters*—18 days at 71°F to 30 days at 64°F.<br>*Olympia Oysters*—30 or more days depending on temperature and food supply. |
| Size of Egg/Larvae | About 0.01 inch at settlement. |
| Growth Rate | *Pacific Oysters*—4 inches in 2–3 years in South Sound; 4–6 years in colder waters. Growth rate is much slower when the oysters are crowded or exposed at low tides.<br>*Olympia Oysters*—Reach marketable size of 1¼ inch in 4–5 years in Puget Sound. |
| Food | Larvae eat phytoplankton; juveniles and adults eat phytoplankton and detritus. Some food is taken in by direct absorption. |

CLAMS
Physical and Biological Characteristics

**Distribution and Zonation**

Distribution

*Butter Clams*—Humboldt Bay, California, to Alaska; a similar species is found south to Baja, California. Common throughout Puget Sound; most abundant in the Strait of Juan de Fuca and central Puget Sound.

*Littleneck Clams*—Baja Peninsula to the Aleutian Islands, Alaska; common throughout Puget Sound. Most abundant in the Strait of Juan de Fuca and central Puget Sound.

*Manila Clams*—Native to Japan and introduced to the U.S. Found in coastal bays and estuaries of British Columbia, Washington, Oregon, and California. Most common in southern Puget Sound.

*Horse Clams*—Northern California to Alaska and common throughout Puget Sound. Usually less abundant than butter clams or native littleneck clams.

*Softshell Clams*—Native to the Atlantic coast from Canada to Florida; introduced to the Pacific Coast from Alaska to San Francisco Bay. Moderately common in Puget Sound.

*Heart Cockles*—Southern California to the Bering Sea. Moderately abundant throughout Puget Sound.

*Geoduck Clams*—Baja Peninsula to Alaska and abundant in Puget Sound.

Zonation

*Butter Clams*—Most common between -3–-30 feet (MLLW); found to depths of at least 180 feet. Mature clams buried up to one foot in the substrate.

*Native Littleneck Clams*—Occur from about the 0 tide zone (MLLW) to a depth of at least 65 feet; most abundant from 0-10 feet; buried up to 6 inches in the substrate.

*Manila Clams*—Most abundant from 4–-2 feet (MLLW); rare subtidally; usually buried close to the surface, to a maximum of 6 inches.

*Horse Clams*—Most comon from −1 foot (MLLW); subtidal depths of 50–60 feet; buried up to 18 inches.

*Softshell Clams*—Most common in high midtidal (6–2 feet MLLW) locations near river mouths and in the less saline portions of bays; usually buried about 8 inches.

*Heart Cockles*—Found from 4 feet in the intertidal zone to subtidal waters from 50–60 feet in depth; usually buried just beneath the surface and may be exposed.

*Geoduck Clams*—Found from the lower intertidal (-2 feet MLLW) zone to depths of at least 200 feet; abundant in certain locations at depths of 30–60 feet; buried to depths of 3 feet.

Substrate

*Butter Clams*—Pure sand to pure gravel; prefers a mixture of sand, broken shell, and small gravel.

*Native Littleneck Clams*—Mixed, firm gravel, cobble and shell; uncommon in sand or mud.

*Manila Clams*—Mixtures of gravel, coarse sand, and small amounts of mud and shell. Does best on beaches with slopes less than 7–8°.

*Horse Clams*—*T. nuttalli.* Pure sand/mud and gravel to medium-sized gravel. *T. capax:* shell and dense sand.

*Softshell Clams*—Mixtures of sand and mud or gravel and mud.

*Heart Cockles*—Mixed sand and mud, common in beds of eelgrass or *Zostera.*

*Geoduck Clams*—Soft mud to sand-gravel mixtures. Most common in mud and sand mixtures of medium compactness.

---

**Physical/Chemical Tolerances**

Temperature

Tolerate from near freezing to 75°F. Except for species with very shallow burrows, clams are not usually exposed to extremes in air or water temperatures. The Manila clam appears fairly tolerant to temperature extremes that may occur on higher intertidal beds during low tides; temperature tolerance of all clams is less during spawning and early development.

Salinity

Optimum salinities are generally 20–35 ppt—clams can tolerate low salinity by closing shells; Manila clams have greater tolerances; softshell clams prefer low salinity waters, as low as 4–5 ppt.

Dissolved Oxygen

Can tolerate low oxygen levels (to 1 mg/liter); require higher levels for sustained growth; withstand very low levels for short periods of time by closing shells.

pH, Ammonia, $H_2S$

A pH range for normal growth of 6.8–8.50; can tolerate high ammonia levels (100–900 ppm); generally intolerant of $H_2S$. Subtidal clams likely to be less tolerant of all factors than intertidal clams.

Currents

Most abundant in areas with moderate currents and good water exchange.

---

**Reproduction**

Type of Reproduction

Sexes are separate and sperm and eggs are extruded directly into the water.

Natural Spawning Season

*Butter Clams*—Late spring to late summer; spawns when water temperature exceeds 60°F.

*Native Littleneck Clams*—Late spring with some spawning through the summer.

*Manila Clams*—Summer with some spawning late spring through early fall.

*Horse Clams*—*T. nuttalli:* summer. *T. capax:* late winter to early spring. *T. capax* spawns at the seasonal minimum water temperature.

*Softshell Clams*—Midsummer to fall.

*Heart Cockles*—Usually in late spring but may spawn from midspring to late summer.

*Geoduck Clams*—March to June when food resources are adequate.

---

**Growth**

Type of Reproduction

Pelagic eggs, planktonic larvae, early juveniles attached by a byssus to bottom materials, free burrowing from juveniles through adults.

*Butter, Native Littleneck, and Manila Clams*—Usually sexually mature after one year, 1½ inches in length.

*Horse Clams*—Sexually mature after 3 years, 2½ inches in length.

| | |
|---|---|
| | *Softshell Clams*—Sexually mature by 1–1½ inches. |
| | *Heart Cockles*—Sexually mature at 2 years. |
| | *Geoduck Clams*—Sexually mature by 3 years. |
| Time for Egg Development | Approximately 12–20 hours to reach the free-swimming stage. |
| Time for Larval Development | About 18 days at 60°F in culture; may exceed 40 days under suboptimal conditions. |
| Size of Egg/Larvae | At settlement, 0.01 inch; usually burrow at about ⅛–¼ inch. |
| Growth Rate | *Butter Clams*—2½ inches in 4–5 years in southern British Columbia, 8–9 years for same size in southeastern Alaska. |
| | *Native Littleneck Clams*—1½ inches in 3½–4 years in Strait of Georgia; maximum size of 2½ inches reached in 10 years. |
| | *Manila Clams*—1½ inches in two growing seasons in Washington or 3½–4 years in Strait of Georgia; maximum size of 2½ inches in 10 years. Growth rate is slower for those higher in intertidal zone. |
| | *Horse Clams*—Reach 5 inches in 4–6 years. |
| | *Softshell Clams*—Grow to about 3 inches in 3 years. |
| | *Heart Cockles*—Sexually mature at 2 years. |
| | *Geoduck Clams*—Usually reach 2 pounds in 8–10 years in Puget Sound, but in some areas reach 2 pounds in 3 years. Achieve maximum weight of 10 pounds. |
| Food | Filter feeders—Consume small plankton, benthic diatoms, and detritus. Most probably eat greater proportions of detrital material. |

MUSSELS
Physical and Biological Characteristics

| | |
|---|---|
| **Distribution and Zonation** | |
| Distribution | *Blue Mussels*—Common throughout Puget Sound and have a worldwide distribution. They are found in estuaries and sheltered waters. |
| | *California Mussels*—In Puget Sound found mainly on exposed areas of the Strait of Juan de Fuca and the San Juan Islands. |
| Zonation | *Blue Mussels*—High to mid intertidal zone; occasionally occur to depths of 120 feet. |
| | *California Mussels*—usually restricted to mid-intertidal zone but may be found down to upper subtidal levels. |
| Substrate | Marine algae, rocks, rope, piling, and other shell including mature mussel shell; natural beds often are high on piling and rocks and are exposed to the air at low tide. |

| | |
|---|---|
| **Physical/Chemical Tolerances** | |
| Temperature | *Blue Mussels*—Tolerate 35—80°F; prefer 50–70°F.<br>*California Mussels*—Optimum is 60–70°F; growth reduced above 70°F. |
| Salinity | Similar to Pacific oysters—optimum is 10–20 ppt. |
| Dissolved Oxygen | Similar to Pacific oysters. |
| pH, Ammonia, $H_2S$ | Similar to Pacific oysters. |
| Currents | *Blue Mussels*—Prefer slow to medium water activity.<br>*California Mussels*—Prefer strong water activity. |

| | |
|---|---|
| **Reproduction** | |
| Type of Reproduction | Separate sexes, external fertilization; reproducing stocks are unevenly distributed in Puget Sound. |
| Natural Spawning Season | *Blue Mussels*—Spawn in late spring through midsummer; duration of spawning is a few weeks for any single area during the season.<br>*California Mussels*—Spawning peak is in July with some spawning in December. |
| Environmental Limits | Spawning is stimulated by increasing water temperature, mechanical action, strong wave action, lunar cycle, and various chemicals. |

| | |
|---|---|
| **Growth** | |
| Type of Reproduction | Pelagic eggs, planktonic larvae, may delay settlement until a suitable bottom type is found; attach to substrate by a byssus and remain attached through adult stage. Are gregarious and tend to form crowded and dense settlements. Sexually mature in one year. |
| Time for Egg Development | About 48 hours. |
| Time for Larval Development | 3–4 weeks is normal, but may remain in plankton for up to 10 weeks. |
| Size of Egg | About .01 inch in diameter. |
| Growth Rate | Varies with food abundance, degree of crowding and exposure of intertidal populations; when constantly submerged may grow to 3–3½ in 2 years; those exposed to air may require 3–4 years to reach 2 inches. |
| Food | Similar to Pacific oysters. |

OCTOPUS
Physical and Biological Characteristics

| | |
|---|---|
| **Distribution and Zonation** | |
| Distribution | Throughout Puget Sound, on both western and eastern North Pacific coasts, as far south as northern California. |
| Zonation | Low intertidal to depths of at least 600 feet; larvae float in mid-water at depths of less than 1,200 feet. |
| Substrate | Prefers crevices and caves in solid outcrops and beneath boulders or manmade debris. |
| **Physical/Chemical Tolerances** | Similar to abalones except the octopus is more common in protected and/or deeper waters. |
| **Reproduction** | |
| Type of Reproduction | Sexes are separate and stable; oviparous with internal fertilization; females brood the eggs; courtship behavior is complex. |
| Natural Spawning Season | Breed from October to December; spawn the following May to July with some spawning occurring either earlier or later in the year. |
| **Growth** | |
| Type of Reproduction | Develop in the egg through well-formed larvae; short pelagic period; benthic from juvenile through adult; mature after 1½–2 years. |
| Time for Egg Development | 6½ months after breeding. |
| Time for Larval Development | A few weeks to months, dependent on water temperature and food supply. |
| Growth Rate | May reach 2 pounds in 12 months and 20 pounds at the end of 2 years. Life span is 3–5 years. |
| Food | Opportunistic carnivores; prefer crab, shrimp, and small benthic fish but may also eat sea cucumbers and starfish. |

SCALLOPS
Physical and Biological Characteristics

| | |
|---|---|
| **Distribution and Zonation** | |
| Distribution | *Weathervane Scallops*—North Puget Sound, San Juan Islands, and Strait of Juan de Fuca.<br>*Rock Scallops*—Throughout Puget Sound, most common in north and central Sound.<br>*Pink Scallops*—Throughout Puget Sound, most common in San Juans and Hood Canal. |
| Zonation | *Weathervane Scallops*—Subtidal from 15–450 feet in depth.<br>*Rock Scallops*—Lower intertidal to subtidal depths of 100–150 feet.<br>*Pink Scallops*—Intertidal to subtidal depths of at least 600 feet. |
| Substrate | *Weathervane and Pink Scallops*—Fine to coarse sand with some resting on solid rock.<br>*Rock scallops*—Attached to solid rock, pilings, etc. |
| **Physical/Chemical Tolerances** | |
| Temperature | Will feed and grow from 50–80°F. |
| Salinity | Cannot tolerate low salinity (below 23 ppt). |
| Dissolved Oxygen | Generally intolerant of low dissolved oxygen concentrations. |
| Currents | Prefer areas with moderate to strong currents. |
| pH, Ammonia, $H_2S$ | Sensitive to moderate fluctuations in these water quality factors; high levels of ammonia depress respiration. |
| **Reproduction** | |
| Type of Reproduction | Sexes are separate and fertilization is external. |
| Natural Spawning Season | Late spring or early fall; exact time of spawning varies. |
| Environmental Limits | Spawning period probably related to changes in water temperature or the presence of natural algae blooms. |
| **Growth** | |
| Type of Reproduction | Pelagic eggs and planktonic larval stage; juveniles attached to a solid or stable substrate by a byssus. Rock scallop permanently cemented when 1 inch in length; weathervane and pink scallop adults not attached and may swim short distances. |
| Time for Larval Development | About 4 weeks, time depends on water temperature and food. |
| Growth Rate | *Weathervane Scallops*—Reach 5 inches in 4 years; maximum size of 9 inches.<br>*Rock Scallops*—Reach up to 4 inches in 3 years. Maximum size varies.<br>*Pink Scallops*—Reach a maximum size of 2½ inches. |
| Food | Similar to oysters and clams. |

ABALONE
Physical and Biological Characteristics

| | |
|---|---|
| **Distribution and Zonation** | |
| Distribution | *Pinto Abalone*—Found from Southeastern Alaska to Southern California; in Puget Sound are most common in Haro Strait.<br>*Red Abalone*—Has a natural range from southern Oregon to central Baja Peninsula; have been transplanted to the San Juan Islands. |
| Zonation | Most common in depths from extreme low water to 100 feet in depth. |
| Substrate | Solid rock or boulders separated by gravel in areas with brown algae or kelp; protective habitats are preferred. |
| **Physical/Chemical Tolerances** | |
| Temperature | Survive and grow from 45–80°F; larvae do poorly above 70°F. |
| Salinity | Require oceanic conditions. |
| Dissolved Oxygen | Tolerate low (1 ppm) levels for only short periods of time. |
| Currents | Prefer areas with moderate to strong currents and medium to high wave activity. |
| pH, Ammonia, $H_2S$ | Probably similar to scallops. |
| **Reproduction** | |
| Type of Reproduction | Sexes with external ferilization; may be induced to spawn by changing the water temperature or exposing them to ultraviolet light-irradiated sea water. |
| Natural Spawning Season | *Pinto Abalone*—Spawns from April through June in British Columbia and Puget Sound.<br>*Red Abalone*—Reported to spawn year round in California. |
| Environmental Limits | Temperature has a marked influence on the rate of growth and development; in culture, short term increases in water temperature induce spawning. |
| **Growth** | |
| Type of Reproduction | Pelagic eggs; eggs develop into ciliated larvae; settle as creeping post-larvae which are able to swim; flattened shell is developed very soon after settlement and pores are formed as the abalone increases in size. |
| Time for Egg Development | 48–72 hours from fertilization to ciliated larvae. |
| Time for Larval Development | *Pinto Abalone*—8–9 days in Puget Sound waters. |
| Growth Rate | *Pinto Abalone*—1 inch in first year and 2¼ inches in 3 years in culture in Puget Sound; up to 3½ inches in 11 years for pinto abalone in British Columbia.<br>*Red Abalone*—In California reach a maximum of 7–8 inches in 6 years. |
| Food | Adults and juveniles prefer brown algae; very young abalone eat diatoms and sometimes red algae as well as green algae. |

DUNGENESS CRABS
Physical and Biological Characteristics

| | |
|---|---|
| **Distribution and Zonation** | |
| Distribution | Found from Alaska to northern Mexico. In Puget Sound, most abundant in north. |
| Zonation | Low intertidal to deep subtidal; most abundant at depths of about 100 feet. |
| Substrate | Sand and mud; juveniles and young found in shallow seaweed and eelgrass beds. |
| **Physical–Chemical Tolerances** | |
| Temperature | Optimum temperature for larvae is 50–60°F; adults tolerate near freezing to 75°F. Crabs approaching molt are less resistant to increased temperature. |
| Salinity | To 9–10 ppt for juvenile and young crabs; larval survival best at 25–30 ppt; intolerant of rapid fluctuations. |
| Dissolved Oxygen | Intolerant of low dissolved oxygen concentrations; optimal is above 5 ppm. |
| Currents | Prefer areas of low to moderate bottom currents. |
| pH, Ammonia, $H_2S$ | Ammonia is toxic at low concentrations. |
| **Reproduction** | |
| Type of Reproduction | Males polygamous; sperm transferred to receptacles of female just after molt; eggs are fertilized later in the receptacles; eggs remain attached to the female until hatching. |
| Natural Spawning Season | Egg-bearing females found every month; most common November–March; mating is usually in late spring or early summer. |
| **Growth** | |
| Type of Reproduction | 6 planktonic larval stages; up to 12 juvenile molts on the bottom to reach-maturity; and up to 14 molts before they reach minimum legal size; usually molt once a year when mature. |
| Time for Egg Development | 7–10 months; hatch in northern Puget Sound mid-March–April. |
| Time for Larval Development | 3–5 months. |
| Growth Rate | Reach a carapace width of about 4 inches in 2 years; first harvested at about 4 years and 6¼ inches. Young crabs may grow up to 1 inch at each molt; after maturity males grow faster than females with larger maximum sizes; maximum size is about 9 inches across the back. |
| Food | Juveniles and adults are opportunistic carnivores; prefer other crustaceans, bivalve molluscs, worms, eggs, and small fish. |

SHRIMP
Physical and Biological Characteristics

**Distribution and Zonation**

| | |
|---|---|
| Distribution | Found throughout Puget Sound, coastal waters of Washington, Alaska, Oregon and northern California. |
| Zonation | Semibenthic as adults. |

**Physical/Chemical Tolerances**

| | |
|---|---|
| Temperature | Optimal from lows of 30°F to 60°F; cannot withstand high water temperatures. |
| Salinity | 25–30 ppt for pink shrimp; larvae more adapted to varying salinity than the adults; low salinity is growth inhibiting; effects more critical than temperature. |
| Dissolved Oxygen | Not known for pandalids; for penaeiids the lower limit is about 5 ppm. |
| Currents/Light | Larvae are attracted to light; juveniles and adults avoid lighted areas; adults leave the bottom in late afternoon and return by dawn. |
| pH, Ammonia, $H_2S$ | Ammonia is toxic at low concentrations; $H_2S$ with low dissolved oxygen may lead to "black gill" disease. |

**Reproduction**

| | |
|---|---|
| Type of Reproduction | Protandric hermaphrodites; usually mature first as males, breed at least once, and then transform to females; function as females through remainder of life. Males transfer sperm to receptacles of females; eggs are fertilized externally at a later date; eggs remain attached to the females until they hatch. |
| Natural Spawning Season | Mate in the fall; eggs are extruded shortly after mating and fertilized. |

**Growth**

| | |
|---|---|
| Type of Reproduction | Eggs attached to female until they hatch; pass through 6 planktonic larval stages. Most Puget Sound shrimp mature during their second year and function one or two years as males before transforming to females: age at maturity varies with species and location. |
| Time for Egg Development | 4–8 months. |
| Time for Larval Development | 2–3 months. |
| Growth Rate | *Coonstripe Shrimp*—Reach carapace lengths ranging from ½–1 inch in their first year. In their second year they range from 1–1⅛ inch in carapace length.<br>*Spot shrimp*—In Hood Canal, reach average carapace lengths of 1 inch (age 1), 1¼ inch (age 2), and 1½ inch (age 3 or 4). |
| Food | Carnivorous bottom feeders; found feeding on marine worms and small crustaceans, detritus, and dead animal matter. |

SEA URCHINS AND SEA CUCUMBERS (*Stichopus*)
Physical and Biological Characteristics

| | |
|---|---|
| **Distribution and Zonation** | |
| Distribution | *Red Sea Urchin*—Most abundant in the Strait of Juan de Fuca, Strait of Georgia, and San Juans; more common in southerly and warmer waters.<br>*Sea Cucumbers*—Common in parts of the San Juans, South Sound, and Whidbey basin. |
| Zonation | *Red Sea Urchin*—Intertidal to depths of 375 feet; tend to cluster in crevices and between kelp holdfasts, especially in high subtidal zone.<br>*Sea Cucumbers*—Prefer eelgrass beds, common from shoreline to depths of 250 feet. |
| Substrate | *Red Sea Urchins*—Prefer rocky bottoms in areas of brown seaweed.<br>*Sea Cucumbers*—Associated with varied bottom types, including solid rock, medium to coarse sand, mixed sand, and eelgrass. Prefers sand/eelgrass substrates. |
| **Physical/Chemical Tolerances** | |
| Temperature | *Red Sea Urchin*—Optimum is slightly higher than the normal Puget Sound range.<br>*Sea Cucumbers*—Range is within Puget Sound values. |
| Salinities | Intolerant of low salinities; prefer oceanic conditions. |
| Dissolved Oxygen | Generally intolerant of low dissolved oxygen concentrations. |
| Currents | *Sea Urchins*—Prefer moderate to swift currents.<br>*Sea Cucumbers*—Usually found in areas with low current velocities. |
| pH, Ammonia, $H_2S$ | Similar to oysters. |
| **Reproduction** | |
| Type of Reproduction | *Sea Urchins*—Sexes are separate with external fertilization; may aggregate prior to spawning.<br>*Sea Cucumbers*—The sexes are separate with external fertilization. |
| Natural Spawning Season | *Red Sea Urchins*—Spawn April through June, but spawning may be sporadic from year to year.<br>*Sea Cucumbers*—Spawn July to August in Puget Sound and September in the San Juans. |
| **Growth** | |
| Type of Reproduction | Planktonic larvae go through a number of stages and metamorphose quickly after settling on the bottom as miniature urchins or sea cucumbers. |
| Time for Larval Development | *Sea Urchins*—2–3 months after spawning.<br>*Sea Cucumbers*—15–23 days. |

| | |
|---|---|
| Size of Egg/Larvae | Eggs—80–90 mm; larvae just before metamorphosis—about 1 mm. |
| Growth Rate | *Sea Urchins*—Species similar to red urchin reach 1½–2¼ inches in diameter in 4 years.<br>*Sea Cucumbers*—Probably grow at a rate similar to urchins. |
| Food | *Sea Urchins*—Grazing herbivores; prefer brown algae or kelp.<br>*Sea Cucumbers*—*Stichopus* is a deposit feeder and lives on organic material and algae that are on the surface or mixed with sand. |

# Appendix B

## Physical and environmental features of commercially harvested pandalid shrimp in Puget Sound

| Size in P.S. | Relative Abundance | Preferred Substrate | Depth Range, ft. (Adults) | Areas of Concentration | Comments |
|---|---|---|---|---|---|
| Spot shrimp *Pandalus platyceros* | | | | | |
| Large (10/lb) | Common | Rock-mud | 12-1500 | Central & south Hood Canal, Lopez Island, Elliott Bay, Bellingham Bay, Possession Sound | Largest shrimp in Puget Sound; most important species in Hood Canal fishery. Has good aquaculture potential. |
| Northern pink shrimp *Pandalus borealis* | | | | | |
| Med-Sm (60/lb) | Frequent | Mud | 180-2100 | Bellingham Bay, San Juan Island, Port Susan, Possession Sound, Carr Inlet, Hood Canal | Dominates the small shrimp landings in Puget Sound. Most important species in Alaska and North Atlantic shrimp fisheries. |
| Ocean pink shrimp *Pandalus jordani* | | | | | |
| Small (90/lb) | Frequent | Silty mud-sand | 150-600 | South Puget Sound from Vashon Island south, Possession Sound, Saratoga Passage | Most abundant off the open coast of Washington. |
| Coonstripe shrimp *Pandalus hypisinotus* | | | | | |
| Med. (30/lb) | Frequent | Rock-mud | 18-470 | Hood Canal, Carr Inlet, Lopez Island, Skagit Bay, Bellingham Bay | Landings are lumped with dock and humpy shrimp. This is probably a minor species in the fishery. |
| Dock shrimp *Pandalus danae* | | | | | |
| Med-Sm (55/lb) | Occasional | Sand-gravel | 60-600 | Mouth of Budd Inlet, Port Orchard, San Juan Island, Discovery Bay, Hood Canal | Often taken in pots; common around piers and docks. |
| Humpy shrimp *Pandalus goniurus* | | | | | |
| Small (80/lb) | Occasional | Sand-gravel | 18-600 | San Juan Islands | Often found in association with dock shrimp. |
| Sidestripe shrimp *Pandalopsis dispar* | | | | | |
| Med-Lrg (20/lb) | Occasional | Soft mud | 150-2100 | Bellingham Bay, San Juan Island, occasionally in Hood Canal | Rarely taken in pots; usually trawl harvested at depths below 250 ft. |

# Appendix C

## Actual or potential uses of marine algae found in Washington

| Scientific Name* | Uses | | | | | Distribution | | | Habitat | | |
|---|---|---|---|---|---|---|---|---|---|---|---|
| | FOOD | FERTILIZER | FODDER | GUMS | CHEMICALS | ST. JUAN DE FUCA | SAN JUAN IS. | PUGET SOUND | SUBTIDAL | LOW INTER | MID–HIGH |
| **Chlorophyta** | | | | | | | | | | | |
| *Chaetomorpha* (3 species) | • | | | | | • | • | • | • | • | |
| *Codium fragile* (Sur.) Harv. (fleece, sponge tang) | • | | | | | • | • | | • | • | |
| *Entermorpha clathrata* (Roth) Grev. var. *clathrata* | • | | | | | | | • | | | • |
| *Enteromorpha compressa* (L.) Grev. | • | | • | | | • | • | | | • | • |
| *Enteromorpha intestinalis* (L.) Link var. *intestinalis (green nori)* | • | • | • | | | • | • | • | | | • |
| *Entermorpha linza* (L.) J. Ag. | • | | • | | | • | • | • | | • | |
| *Entermorpha prolifera* (Mueller) J.Ag. | • | | | | | • | • | • | | • | |
| *Monostroma* (3 species) ("awo-nori") | • | | • | | | • | • | • | | • | • |
| *Prasiola* sp. (1 species; different from one used for food) | • | | | | | • | • | | | | • |
| *Ulva lactuca* (L.) (Sea lettuce) | • | | | | | • | • | • | • | • | • |

*Scientific names follow Abbott and Hollenberg, 1976.

Actual or potential uses of marine algae found in Washington

| Scientific Name* | Uses | | | | | Distribution | | | Habitat | | |
|---|---|---|---|---|---|---|---|---|---|---|---|
| **Phaeophyta** | FOOD | FERTILIZER | FODDER | GUMS | CHEMICALS | ST. JUAN DE FUCA | SAN JUAN IS. | PUGET SOUND | SUBTIDAL | LOW INTER | MID-HIGH |
| *Agarum cribosum* Bory | | | | • | | • | • | | • | | |
| *Agarum fimbriatum* Harv. | | | | • | | • | • | | • | | |
| *Alaria marginata* P. *et* R. (wing kelp) | • | • | | • | | • | • | | | • | |
| *Analipus japonicus* (Harv.) Wynne (fir needle) | • | | | | | • | • | | | | • |
| *Costaria costata* (C.Ag.) Saunders | | | | • | | • | • | • | • | | |
| *Cymathere triplicata* (P. *et* R.) J. Ag. | | | | • | | • | • | | • | | |
| *Egregia menziesii* (Turner) Aresch. | | • | | • | | • | • | | • | | |
| *Fucus distichus* subsp. *edentatus* (de la Pyl.) Powell (Rockweed) | | • | | | | • | • | • | | | • |
| *Laminaria* (8 species) | • | • | • | • | | • | • | • | • | | |
| *Laminaria saccharina* (L.) Lam. ("Konbu") | • | | | | | • | • | • | • | | |
| *Macrocystis integrifolia* Bory (Giant kelp) | | • | • | • | | • | | | • | | |
| *Nereocystis luetkeana* (Mert.) P. *et* R. (Bull kelp) | • | • | • | • | | • | • | • | • | | |
| *Pleurophycus gardneri* Setchell *et* Saunders (kelp) | • | | | | | • | • | | • | | |
| *Postelsia palmaeformis* Rupr. (Sea palm) | • | | | | | • | | | | | • |
| *Scytosiphon lomentaria* (Lynb.) J. Ag. | • | | | | | • | • | • | • | • | |

*Scientific names follow Abbott and Hollenberg, 1976.

Actual or potential uses of marine algae found in Washington

| Scientific Name* | Uses | | | | | Distribution | | | Habitat | | |
|---|---|---|---|---|---|---|---|---|---|---|---|
| | FOOD | FERTILIZER | FODDER | GUMS | CHEMICALS | ST. JUAN DE FUCA | SAN JUAN IS. | PUGET SOUND | SUBTIDAL | LOW INTER | MID–HIGH |
| **Rhodophyta** | | | | | | | | | | | |
| *Ahnfeltia gigartinoides* J. Ag. | • | | | • | | • | • | | | • | |
| *Bangia fuscopurpurea* (Dillwyn) Lynb. (cow hair) | • | | | | | • | • | • | | | • |
| *Ceramium pacificum* (Coll.) Kylin | | | | • | | • | • | • | • | • | |
| *Chondrus ocellatus* f. *parvus* Mikami | | | | • | | • | | | | | • |
| *Constantinea simplex* Setchell | | | | | | • | • | • | • | | |
| *Corallina officinalis* var. *chilensis* (Dec.) Kutz. | VERMIFUGE | | | | | • | • | | | • | |
| *Cryptosiphonia woodii* (J. Ag.) J. Ag. | ANTIVIRAL | | | | • | • | | | | • | |
| *Endocladia muricata* (P. *et* R.) J. Ag. | | | | • | | • | • | | | | • |
| *Farlowia compressa* J. Ag. f. *mollis* (Harv. *et* Bail.) Farl. *et* Setch. | | | | | | • | • | | • | • | |
| *Gelidium* (6 species) | | | | • | | • | • | | • | • | |
| *Gigartina* (11 species) | | | | • | | • | • | • | • | • | • |
| *Gloiopeltis furcata* (P. *et* R.) J. Ag. ("funori") | • | | | • | | • | • | | | | • |
| *Gracilaria verrucosa* (Hudson) Papenfuss ("ogo") | | | | • | | • | • | • | • | • | |
| *Gracilariopsis sjoestedtii* (Kylin) Dawson | | | | • | | • | • | • | • | • | |
| *Grateloupia* (3 species) | • | | | • | | • | • | • | • | • | |
| *Gymnogongrus* (4 species) | | | | • | | • | • | | • | • | |

*Scientific names follow Abbott and Hollenberg, 1976.

Actual or potential uses of marine algae found in Washington

| Scientific Name* | Uses | | | | | Distribution | | | Habitat | | |
|---|---|---|---|---|---|---|---|---|---|---|---|
| | FOOD | FERTILIZER | FODDER | GUMS | CHEMICALS | ST. JUAN DE FUCA | SAN JUAN IS. | PUGET SOUND | SUBTIDAL | LOW INTER | MID-HIGH |
| **Rhodophyta (cont.)** | | | | | | | | | | | |
| *Halosaccion glandiforme* (Gmelin) Ruprecht (sea sac) | • | | | | | • | • | | | • | • |
| *Iridaea* (8 species) | | | | • | | • | • | • | • | • | • |
| *Laurencia* (1 species, 2 varieties) | • | | | | • | • | • | | | • | |
| *Lithothamnion* (4 species) | | • | | | | • | • | | • | | |
| *Nemalion helminthoides* (Vell.) Batt. (sea noodle) | • | | | | | • | | | | | • |
| *Neoagardhiella baileyi* (Kurtzt.) Wynne *et* Taylor | • | | | • | | • | • | • | • | | |
| *Neodilsea americana* Abbott | | | | • | • | • | • | | • | | |
| *Odonthalia floccosa* (Esper.) Falkenb. | BROMINE | | | | | • | • | • | | • | |
| *Palmaria palmata* f. *mollis* (S. *et* G.) Guiry (Dulse) | • | | | | | • | • | | • | • | |
| *Pikea californica* Harv. | ANTIVIRAL | | | | | • | | | • | • | |
| *Polyneura latissima* (Harv.) Kylin | • | | | | | • | • | • | • | | |
| *Porphyra abbottae* Krish. ("Nori") | • | | | | | • | • | • | • | • | |
| *Porphyra nereocystis* Anderson ("Nori") | • | | | | | • | • | | • | | |
| *Porphyra perforata* J. Ag. ("Nori") | • | | | | • | • | • | • | | | • |
| *Porphyra pseudolanceolata* Krish. ("Nori") | • | | | | | • | • | | | | • |
| *Porphyra torta* Krish. ("Nori") | • | | | | | • | • | | | | • |
| *Rhodoglossum* (4 species) | | | | • | | • | | | | | • |
| *Rhodomela larix* (Turner) C. Ag. | | | | • | | • | • | | | • | • |

*Scientific names follow Abbott and Hollenberg, 1976.

**28B.20.320 Marine biological preserve—Established and described.** There is hereby created an area of preserve of marine biological materials useful for scientific purposes, except when gathered for human food, and except, also the plant nereocystis, commonly called "kelp." Said area of preserve shall consist of the salt waters and the beds and shores of the islands constituting San Juan county and of Cypress Island in Skagit county (1969 ex.s. c 223 § 28B.20.320. Prior: 1923 c 74 § 1;RRS § 8436-1. Formerly RCW 28.77.230.)

**28B.20.322 Marine biological preserve—Gathering permit.** No person shall gather said marine biological materials from said area of preserve, except upon permission first granted by the director of the Friday Harbor Laboratories of the University of Washington. (1969 ex.s. c. 223 § 28B.20.322. Prior: 1923 c 74 § 2; RRS § 8436-2. Formerly RCW 28.77.231, 28.77.230, part.)

**28B.20.324 Marine biological preserve—Penalty for unlawful gathering.** Any person gathering said marine biological materials contrary to the terms of RCW 28B.20.320 and 28B.20.322 shall be guilty of a misdemeanor. (1969 ex.s. c223 § 28B.20.324. Prior: 1923 c 74 § 3; RRS § 8436-3. Formerly RCW 28.77.232, 28.77.230, part.)

**WAC 322-30-154 Marine aquatic plant removal (RCW 79.68.080).** (1) Any species of aquatic plant may be collected from aquatic land for educational, scientific and personal purposes up to 50 pounds wet weight per year, except that no annual species can be collected in excess of fifty percent of its population's total wet weight in any 1 acre area or any perennial in excess of seventy-five percent of its population's total wet weight in any 1 acre area.

(2) Aquatic plants listed on the commercial species list may be collected without a permit from aquatic land for commercial purposes up to the limits noted on the list, except that no annual species can be collected in excess of fifty percent of its population's total wet weight in any 1 acre area or any perennial in excess of seventy-five percent of its population's total wet weight in any 1 acre.

(3) Aquatic plants may be collected from aquatic land for educational, scientific or personal purposes beyond the weight limitations stated in subsection (1) only through benefit of an aquatic plant removal permit from the department. Payment of a royalty dependent on species, volume and use shall be a condition of the permit.

(4) Aquatic plants as listed on the commercial species list may be collected from aquatic land for commercial purposes beyond the eight limitations stated in subsection (2) only through benefit of an aquatic plant removal permit from the department. Payment of a royalty dependent on species, volume and use shall be a condition of the permit.

(5) Aquatic plants may not be removed from the San Juan Marine Reserve except as provided for in RCW 28B.20.320 and from other areas where prohibited.

(6) Removal of perennial plants must be in such a manner as to maintain their regeneration capability at the site from which thay have been collected.

(7) Species may be deleted or added to the commercial species list through petition to the department.

(8) Species not on the commercial species list may be collected for purposes of market testing, product development, or personal use through either written authorization from the department or through an aquatic plant removal permit depending on the amount of plant material required.

(9) Commercial species list.

| **Species Name** | **Maximum Free Collection Weight** |
|---|---|
| *Alaria marginata* Post. *et* Rupr | 50 pounds wet weight |
| *Cymathere triplicata* (Post *et* Rupr) J.Ag. | 50 pounds wet weight |
| *Gracilaria sjoestedtii* Kylin | 10 pounds wet weight |
| *Gracilaria verrucosa*(Huds) Papenf. | 10 pounds wet weight |
| *Iridaea cordata* (Turner) Bory | 50 pounds wet weight |
| *Laminaria dentigera* (Kjellm.) (*L. setchellii Silva*) | 50 pounds wet weight |
| *Laminaria saccharina* (L.) Lamourous | 100 pounds wet weight |
| *Macrocystis integrifolia Bory* | 100 pounds wet weight |
| *Monostroma* spp. | 20 pounds wet weight |
| *Neoagardhiella baileyi* (Harvey *et* Kutzing) Wynne *et* Taylor | 30 pounds wet weight |
| *Porphyra* spp. | 10 pounds wet weight |
| *Ulva* spp. | 20 pounds wet weight |

(10) Harvesting of fishery resources adhering to marine aquatic plants, such as fish eggs, must be according to the law and as specified by the department of fisheries. A permit may also be required according to WAC 332-30-154(4). (Statutory Authority: TCS 43.30.151. 80-09-005 (Order 343), § 332-30-154, filed 7/3/80.)

# References

CHAPTER ONE

Angell, T. and K.C. Balcomb. 1982. *Marine Birds and Mammals of Puget Sound*. Seattle: Washington Sea Grant Program. 145 pp.

Bish, R.L. 1982. *Governing Puget Sound*. Seattle: Washington Sea Grant Program. 137 pp.

Burns, R. 1985. *The Shape and Form of Puget Sound*. Seattle: Washington Sea Grant Program. 100 pp.

Carefoot, T. 1977. *Pacific Seashores*. Seattle: University of Washington Press. 208 pp.

Chasan, D.J. 1981. *The Water Link*. Seattle: Washington Sea Grant Program. 179 pp.

Downing, J. 1983. *The Coast of Puget Sound*. Seattle: Washington Sea Grant Program. 126 pp.

FAO. 1982. Yearbook of fishery statistics, 1981. Volume 52. Food and Agriculture Organization of the United Nations, Rome.

Kozloff, E.N. 1983. *Seashore Life of the Northern Pacific Coast*. Seattle: University of Washington Press.

Nishitani, L. and K.K. Chew. 1984. Recent developments in paralytic shellfish poisoning research. *Aquaculture*. 38:317–329.

Saunders, R.S. 1984. Shellfish protection strategy. Washington Department of Ecology. Olympia, Washington.

Strickland, R.M. 1983. *The Fertile Fjord*. Seattle: Washington Sea Grant Program. 145 pp.

Washington State Department of Fisheries. 1983. 1982 Fisheries Statistical Report. Olympia, Washington.

———. 1978. Washington State Shellfish. Olympia, Washington. 61 pp.

———. 1978. Puget Sound Public Shellfish Sites. Olympia, Washington. 31 pp.

CHAPTER TWO

Beattie, J.H., W.K. Hershberger, K.K. Chew, C. Mahnken, E.F. Prentice and C. Jones. 1978. Breeding for resistance to summertime mortality in the Pacific oyster (*Crassostrea gigas*). Washington Sea Grant Program, University of Washington, Seattle. WSG 78-3. 13 pp.

Brown, B.F. 1977. The role of *Vibrios* as pathogens for the Pacific oyster. M.S. thesis, University of Washington, Seattle.

Chapman, W.M. and A.H. Banner. 1949. Contributions to the life history of the Japanese oyster drill, *Tritonalia japonica*, with notes on other enemies of the Olympia oyster, *Ostrea lurida*. Washington State Department of Fisheries. *Biol. Bull.* 49-A:169–200.

Chew, K.K. 1983. Current status of hatchery-based mariculture in the Pacific Northwest of North America. Fisheries Research Institute, University of Washington, Seattle. Contribution No. 632. 20 pp.

______. 1983. Recent changes in molluscan culture fishery for the Pacific Northwest of the United States. Fisheries Research Institute, University of Washington, Seattle. Contribution No. 633. 24 pp.

______. (ed.). 1982. Proceedings of the North American oyster workshop. World Mariculture Society, Special Publication No. 1.

Esveldt, G.D. 1948. A review of the oyster industry of the State of Washington. South Bend, Washington. Unpublished manuscript.

Glude, J. B. and K. K. Chew. 1982. Shellfish aquaculture in the Pacific Northwest. University of Alaska, Alaska Sea Grant Report 82-2. p. 291–304.

Glude, J.B. 1978. The Pacific oyster. Report prepared for the Electric Power Research Institute and the Tennessee Valley Authority.

Galtsoff, P.S. 1929. Oyster industry of the Pacific coast of the United States. U.S. Department of Commerce, Report of Commissioner of Fisheries. Appendix VIII:367–400.

Gunter, G. and J.E. McKee. 1960. On oysters and sulfite waste liquor. Washington State Pollution Control Commission. 93 pp.

Korringa, P. 1976a. *Farming the Cupped Oyster of the Genus* Crassostrea: *A Multidisciplinary Treatise*. New York: Elsevier Scientific Publishing Co.

______. 1976b. *Farming the Flat Oyster of the Genus* Ostrea: *A Multidisciplinary Treatise*. New York: Elsevier Scientific Publishing Co.

Krantz, G. E. 1982. Oyster hatchery technology series. Cooperative Extension Service. College Park: University of Maryland. 126 p.

Magoon, C. and R. Vining. 1981. Introduction to shellfish aquaculture in the Puget Sound region. Washington Department of Natural Resources. Olympia, Washington.

Mann, R. (ed.) 1979. Exotic species in mariculture. Proc. Symposium Exotic Species Mariculture: Case Histories of the Japanese oyster, *Crassostrea gigas*, with implications for other fisheries. Woods Hole, MA: Woods Hole Oceanographic Institute.

Muir, J. F. and R. J. Roberts (eds.). 1983. *Recent Advances in Aquaculture*. Westview Press. 453 p.

Quayle, D.B. 1971. Pacific oyster raft culture in British Columbia. Fish. Res. Bd. Can. Bull.

______. 1969. Pacific oyster culture in British Columbia. Fish. Res. Bd. Canada. Bull. 169. 193 pp.

Steele, E.N. 1964. *The Immigrant (Pacific) Oyster*. Seattle: Warren's Quick Press.

______. 1957. *The Rise and Decline of the Olympia Oyster*. Elma, WA: Fulco Publications.

Weller, C. 1978. Raft culture of Pacific oysters (*Crassostrea gigas*) in Puget Sound. M.S. Thesis. University of Washington, Seattle.

Westley, R.E. 1978. Past, present and future trends in oyster production and cultural techniques in Washington State. Washington State Department of Fisheries. Prepublication manuscript. Olympia, Washington.

CHAPTER THREE

Anderson, G.J. 1982. Intertidal culture of the Manila clam, *Tapes phillippinarum*, using large netting enclosures in Puget Sound, Washington. M.S. Thesis. University of Washington, Seattle. 100 p.

Anderson, G., M.B. Miller and K.K. Chew. 1982. A guide to Manila clam aquaculture in Puget Sound. Washington Sea Grant Program, University of Washington, Seattle. WSG 82-4. 45 p.

Glock, J.W. 1977. Growth, recovery and movement of Manila clams, *Venerupis japonica*, planted under protective devices and on open beaches and Squaxin Island, Washington. M.S. Thesis, University of Washington, Seattle.

Glude, J.B. 1978. The clams. Report prepared for the Electric Power Research Institute and the Tennessee Valley Authority. 74 pp.

______. 1964. The effect of scoter duck predation on a clam population in Dabob Bay, Washington. *Proceedings of the National Shellfisheries Association*. 55:73–86.

Goodwin, C.L. 1978. Puget Sound subtidal geoduck survey data. Washington State Department of Fisheries. Progress Report 36. 107 pp.

______. 1978. Some effects of subtidal geoduck (*Panope generosa*) harvest on a small experimental plot in Puget Sound, Washington. Washington State Department of Fisheries. Progress Report 66. 21 pp.

______. 1975. Observations on spawning and growth of subtidal geoducks (*Panope generosa* Gould). *Proceedings of the National Shellfisheries Association*, 65:49–58.

______. 1973. Subtidal geoducks of Puget Sound, Washington. Washington State Department of Fisheries Technical Report 13. 64 pp.

______. 1973. Distribution and abundance of subtidal hard-shell clams in Puget Sound, Washington. Washington State Department of Fisheries Technical Report 14. 81 pp.

Goodwin, C.L. and W. Shaul. 1984. Age, recruitment and growth of the geoduck clam (*Panope generosa* Gould) in Puget Sound Washington. Washington Department of Fisheries. Progress Report No. 215. 29 pp.

______. 1981. Puget Sound subtidal geoduck and hardshell clam survey data April 1980–April 1981. Washington State Department of Fisheries. Progress Report No. 215. 29 pp.

______. 1978. Puget Sound subtidal hardshell clam survey data. Washington State Department of Fisheries. Progress Report 44. 92 pp.

______. 1978. Puget Sound hardshell clam survey data March 1977 to March 1978. Washington State Department of Fisheries. Progress Report 64. 12 pp.

______. 1978. Some effects of the mechanical escalator shellfish harvester on a subtidal clam bed in Puget Sound, Washington. Washington State Department of Fisheries. Progress Report 53. 23 pp.

______. 1978. Puget Sound subtidal geoduck survey data March 1977 to March 1978. Washington State Department of Fisheries, Progress Report 65. 30 pp.

Houghton, J.P. 1973. The intertidal ecology of Kiket Island, Washington, with emphasis on age and growth of *Protothaca staminea* and *Saxidomus giganteus*. Ph.D. Thesis, University of Washington, Seattle.

Jones, C.R. 1974. Initial mortality and growth of hatchery-reared Manila clams, *Venerupis japonica*, planted in Puget Sound, Washington beaches. M.S. Thesis, University of Washington, Seattle. 90 pp.

Magoon, C. and R. Vining. 1981. Introduction to shellfish aquaculture in the Puget Sound region. Washington Department of Natural Resources, Olympia, Washington.

Miller, M.B. 1982. Recovery and growth of hatchery-produced juvenile Manila clams *Venerupis japonica* (Deshayes), placed on several beaches in Puget Sound. M.S. Thesis. University of Washington, Seattle. 250 pp.

Mottet, M.G. 1984. Summaries of Japanese papers on hatchery technology and intermediate rearing facilities for clams, scallops and abalones. Washington State Department of Fisheries. Progress Report No. 203. 86 pp.

———. 1980. Research problems concerning the culture of clam spat and seed. Washington Department of Fisheries. Technical Report No. 63.

Oceanographic Commission of Washington. 1981. Clam and mussel harvesting industries in Washington State. Prepared by the Oceanographic Institute of Washington.

Quayle, D.B. and N. Bourne. 1972. The clam fisheries of British Columbia. Fish. Res. Bd. Can. Bull. 179. 70 pp.

Schink, T.D., K.A. McGraw and K.K. Chew. 1983. Pacific coast clam fisheries. Washington Sea Grant Program, University of Washington, Seattle. Technical Report WSG 83-1. 72 pp.

Tarr, M. 1977. Some effects of hydraulic clam harvesting on water quality in Kilisut Harbor, Port Susan, and Agate Pass, Washington. Washington State Department of Fisheries. Progress Report 22. 82 pp.

Vining, R. 1978. Environmental Impact Statement for Commercial Harvest of Clams with a Hydraulic Escalator Dredge. Washington State Department of Natural Resources, Olympia.

Washington Department of Natural Resources and Department of Fisheries. 1981. Management plan for the Puget Sound commercial geoduck fishery. Olympia, Washington.

———. 1981. Management plan for the Puget Sound commercial subtidal hard-shell clam fishery. Olympia, Washington.

Williams, J.G. 1978. The influence of adults on the settlement, growth, and survival of spat in the commercial important clam, *Tapes japonica* Deshayes. M.S. Thesis. University of Washington, Seattle.

CHAPTER FOUR

Beaudry, C.G., Jr. 1984. Survival and growth of larvae of the pinto abalone at different temperatures and culture conditions. M.S. Thesis. University of Washington, Seattle.

Berg, E.R., N.N. FitzSimmons, and T.L. Johnson. 1980. A commercial octopus fishery for the Pacific Northwest and Alaska: Development of prospects. University of Washington, student report for Ocean Engin. Design Course 551. 34 pp.

Bourne, N. 1969. Scallop resources of British Columbia. Fish. Res. Bd. Can. Tech. Rep. No. 104.

Chaves, L. 1975. Mussel culture in Hood Canal. Washington Sea Grant Program, University of Washington, Seattle. WSG TA-75-14.

Cox, K.W. 1960. Review of the abalone in California. *California Fish and Game*. 46(4):381–406.

Falmagne, C.M. 1984. The combined effect of temperature/salinity on survival and growth of *Mytilus californianus* larvae (a response surface analysis). M.S. Thesis. University of Washington, Seattle.

FAO/UNDP South China Sea Programme. 1983. Joint ADB/FAO (SCSP-INFOFISH) market studies. Vol 4: The international market for cephalopods. Manila SCS/DEV/83/24. 53 p.

Fields, W.G. 1965. The structure, development, food relations, reproduction, and life history of the squid *Loligo opalescens* Berry. California State Department of Fish and Game. Fish. Bull. 131. 108 pp.

Fralick and Tillapaugh. 1979. A review of smaller fishery mariculture industries in British Columbia: mussels, abalone, geoduck clams, horse clams and marine plants. Info. Report No. 3. Westwater Research Centre. University of British Columbia.

Hanlon, R.T., R.F. Hixon, W.H. Hulet, and W.T. Yang. 1978. Rearing experi-

ments on the California market squid *Loligo opalescens* Berry, 1911. *The Veliger*. 21(4):428–431.
High, W.L. 1976. The giant Pacific octopus. NOAA, National Marine Fisheries Service, Marine Fisheries Review, Paper 1200:16–22.
Idyll, C.P. 1979. Mussel round-up. *Oceans*. 16–19.
Johnson, K. W. 1979. The relationship between *Mytilus edulis* larvae in the plankton and settlement for Holmes Harbor, Washington. M.S. Thesis, University of Washington, Seattle. 45 pp.
Kjerskog-Agersborg, H.P. 1920. The utilization of echinoderms and of gastropod mollusks. *The American Naturalist*. 54:414–426.
Korringa, P. 1976. *Farming marine organisms low in the food chain: Developments in aquaculture and fisheries science, 1*. New York: Elsevier Scientific Publishing Co.
Leighton, D.L. and C. Phleger. 1981. The suitability of the purple rock scallop to marine aquaculture. San Diego St. University Sea Grant Publication. T-SCSGP-001. 85 pp.
Lutz, R.A. (ed.) 1980. *Mussel Culture and Harvest: A North American Perspective*. New York: Elsevier Scientific Publishing Co. 350 pp.
Mottet, M.G. 1984. Summaries of Japanese papers on hatchery technology and intermediate rearing facilities for clams, scallops and abalones. Washington State Department of Fisheries. Progress Report No. 203. 86 pp.
———. 1978. A review of the fishery biology of abalones. Washington State Department of Fisheries Technical Report 37. 81 pp.
———. 1978. A review of the fishery biology of scallops. Washington State Department of Fisheries. Technical Report 39. 100 pp.
———. 1975. The fishery biology of *Octopus dofleini* (Wulker). Washington State Department of Fisheries Technical Report 16. 39 pp.
Myers, E. A. 1981. The husbandry of mussels in a Maine estuary: An approach to a commercial enterprise. University of New Hampshire, Durham. 45 p.
Oceanographic Commission of Washington. 1981. Clam and mussel harvesting industries in Washington State. Prepared by the Oceanographic Institute of Washington.
Olsen, S. 1984. Completion report on invertebrate aquaculture, shellfish enhancement project 1978–1983. Washington Department of Fisheries, Olympia, Washington.
Peters, J.A. 1978. Scallops and their utilization. NOAA, National Marine Fisheries Service. *Marine Fisheries Review*. 40(11):1–9.
Quayle, D.B. 1971. Growth, morphometry, and breeding in the British Columbia abalone (*Haliotis kamtschatkana* Jonas). Fish. Res. Bd. Can. Tech. Rep. No. 279.
Skidmore, D.A. 1983. Settlement, growth, and survival of *Mytilus edulis* in Puget Sound and assessment of *Mytilus californianus* for aquaculture. M.S. Thesis, University of Washington, Seattle. 99 pp.
Slabyj, B.M., D.L. Creamer, and R.H. True. 1978. Seasonal effect on yield, proximate composition, and quality of blue mussel, *Mytilus edulis*, meats obtained from cultured and natural stocks. NOAA, National Marine Fisheries Service, *Marine Fisheries Review*. 40(8):18–23.
Street, A.D. 1983. Squid fishery development project for Southeast Alaska. Alaska Fisheries Development Foundation, Inc. Anchorage. 65 pp.
Waterstrat, P.R. 1979. Prospects for the development of a mussel culture industry in Puget Sound. M.S. Thesis, University of Washington, Seattle. 60 pp.
Wilson, J.R. and A.H. Gorham. 1982a. Alaska underutilized species. Vol. I: squid. University of Alaska, Alaska Sea Grant Program. Report 82-1. 77 pp.

______. 1982b. Alaska underutilized species. Vol. II: octopus. University of Alaska, Alaska Sea Grant Program. Report 82-3.

CHAPTER FIVE

Armstrong, D.A., B.G. Stevens, and J.C. Hoeman. 1981. Distribution and abundance of Dungeness crab and crangon shrimp and dredging-related mortality of invertebrates and fish in Grays Harbor, Washington. U.S. Army Corps of Engineers, Seattle.

Bumgarner, D. 1978. Puget Sound crab and shrimp. Washington State Department of Fisheries. Progress report for 309 project.

______. 1977. Puget Sound shrimp and crab management study. Washington State Department of Fisheries. Progress report for 309 project.

Butler, T.H. 1960. Maturity and breeding of the Pacific edible crab, *Cancer magister* Dana. *J. Fish. Res. Bd. Can.* 17(5):641–646.

Cleaver, F.C. 1949. Preliminary results of the coastal crab (*Cancer magister*) investigation. Washington State Department of Fisheries. Biol. Bull. 49A:47–82.

Hart, J.F.L. 1982. Crabs and their relatives of British Columbia. Victoria: British Columbia Provincial Museum. 267 pp.

Harville, J.P. and Verhoeven, L.A. 1978. Dungeness crab project of the State-Federal Fisheries Management Program. Portland: Pacific Marine Fisheries Commission.

Hayes, M.L. 1973. Commercial crabs in the eastern north Pacific Ocean. NOAA, National Marine Fisheries Service, Seattle. Unpublished manuscript.

Hopkins, F.W. 1972. Dungeness crab pots. NOAA, National Marine Fisheries Service Extension Publication. Fishery Facts—3, 13 p.

Hoopes, D.T. 1973. The Dungeness crab. NOAA, National Marine Fisheries Service, Extension Publication. Fishery Facts—6.

Mayer, D.L. 1973. The ecology and thermal sensitivity of the Dungeness crab, *Cancer magister*, and related species of the benthic community in Similk Bay, Washington. Ph.D. Dissertation, University of Washington, Seattle. 188 pp.

Reed, P.H. 1969. Culture methods and effects of temperature and salinity on survival and growth of Dungeness crab (*Cancer magister*) larvae in the laboratory. *J. Fish. Res. Bd. Can.* 26:389–397.

Stevens, B.G., D.A. Armstrong and R. Casimano. 1982. Feeding habits of the Dungeness crab *Cancer magister* as determined by the index of relative importance. Mar. Biol. 72:135–145.

Wild, P.W. and R.N. Tasto. 1983. Life history, environment, and mariculture studies of the Dungeness crab, *Cancer magister*, with emphasis on the central California fishery resource. California Department of Fish and Game. Fish Bulletin 172. 352 p.

Williams, J.G. 1975. The intertidal Dungeness crab (*Cancer magister*) sport fishery in Puget Sound. M.S. Thesis, University of Washington, Seattle. 51 pp.

CHAPTER SIX

Anderson, P.J. 1973. Shrimp fishery of the northeastern Pacific Ocean and Bering Sea. NOAA, National Marine Fisheries Service, Kodiak, Alaska. Unpublished manuscript.

Bumgarner, D. 1978. Puget Sound crab and shrimp. Washington State Department of Fisheries. Progress report for 309 project.

Butler, T. H. 1980. Shrimps of the Pacific Coast of Canada. Ottawa: Canada Department of Fisheries and Oceans. 280 pp.

———. 1977. Puget Sound shrimp and crab management study. Washington State Department of Fisheries. Progress report for 309 project.

Fox, W.W. Jr. 1972. Dynamics of exploited pandalid shrimps and an evaluation of management models. Ph.D. Dissertation, University of Washington.

Jones, C.R. 1978. Development of management techniques for a sea urchin fishery. Washington State Department of Fisheries. Progress report for 309 project.

Magoon, C.D. 1979. Shrimp fishing in Washington. Washington State Department of Fisheries. Information Booklet No. 3. 24 p.

Mottet, M.G. 1976. The fishery biology and market preparation of sea cucumbers. Washington State Department of Fisheries. Tech. Rept. 22. 44 pp.

———. 1976. The fishery biology of sea urchins in the family Strongylocentrotidae. Washington State Department of Fisheries. Tech. Rept. 20. 66 pp.

Rensel, J.E. 1976. A comparison of growth and survival of culture spot prawns *Pandalus platyceros* Brandt at two salmon farming sites in Puget Sound. M.S. Thesis, University of Washington, Seattle. 90 pp.

Ronholt, L.L. 1974. A study of the relative efficiency of shrimp pots for harvesting the spot shrimp *Pandalus platyceros*, in southeastern Alaskan waters. M.S. Thesis. University of Washington, Seattle.

Smith, R.T. 1937. Observations on the shrimp fishery in Puget Sound. Washington Department of Fisheries. Biological Report No. 36D.

Wilson, J.R. and A.H. Gorham. 1982. Alaska underutilized species. Vol. III: sea urchins. University of Alaska, Alaska Sea Grant Program. Report 82-7.

CHAPTER SEVEN

Abbott, I.A., and G.J. Hollenberg, 1976. *Marine Algae of California*. Stanford, CA: Stanford University Press. 827 pp.

Abbott, I.A., M.S. Foster, and L.F. Eclund (eds.). 1980. *Pacific Seaweed Aquaculture*. Proc. Symp., March 6–8, 1980. Pacific Grove, California Sea Grant Program, UCSD 80-92, LaJolla, 228 pp.

Adams, R.W., and A. Austin. 1979. Potential yields of *Iridaea cordata* (Florideophyceae) in natural and artificial populations in the Northeast Pacific. *Proceedings of the Ninth International Seaweed Symposium*. 9:499–509.

Allen, G. 1984. *Bioeconomics of Aquaculture*. New York: Elsevier Scientific Publishing Co. 351 pp.

Austin, A. and R. Adams. 1975. Red algal resource studies in Canadian Pacific waters: Carrageenophyte inventory and experimental cultivation phase 1974/1975. *Report to the Federal Ministry of Fisheries and the Provincial Minister of Recreation and Conservation*. Vol. I. Text, Vol. II, Appendices.

Bardach, J.E., J.H. Ryther, and M.O. McLarney. 1972. *Aquaculture: The Farming and Husbandry of Freshwater and Marine Organisms*. New York: Wiley-Interscience. 868 pp.

Bidwell, R.G.S., J. McLachlan, and N.D.H. Llord. 1985. Tank cultivation of Irish Moss, *Chondrus crispus* Stackh. *Botanica Marina* 28:87–97.

Bixler, H.J. 1979. Manufacturing and marketing carrageenan. In: Santilices, B. (ed.) *Actas del Primer symposium sobre Algas Marinas Chilenas*. Subscretario de Pesca, Minist. Econ. Fomento y Recontr. Chile. pp. 259–274.

Blunden, G. 1972. The effects of aqueous seaweed extract as a fertilizer additive. *Proceedings of the Seventh International Seaweed Symposium*. 7:584–589.

Bold, H.C., and M.J. Wynne. 1985. *Introduction to the Algae: Structure and Reproduction* (2nd Ed.). Englewood Cliffs, NJ: Prentice-Hall, Inc. 720 pp.

Boney, A.D. 1965. Aspects of the biology of the seaweeds of economic importance. *Advances in Marine Biology*. 3:105–253.

Booth, E. 1969. The manufacture and properties of liquid seaweed extracts.

*Proceedings of the Sixth International Seaweed Symposium*. 6:655–662.
Brehany, J.J. 1983. An Economic and Systems Assessment of the Concept of Nearshore Kelp Farming for Methane Production, Final Report. Gas Research Institute, Chicago.
Byce, W.J., T.F. Mumford, M. Inayoshi, D.J. Melvin, and V.M. Bryant. 1984. Equipment for Nori Farming in Washington State. Vol. 1, Outdoor Seeding. Vol. 2, Nursery Culture. Department of Natural Resources, Olympia, Washington. 154 pp.
Cameron, F.K. 1915. *Potash from Kelp*. U.S. Department of Agriculture Report No. 10.
Carefoot, T. 1977. *Pacific Seashores: A Guide to Intertidal Ecology*. Seattle: University of Washington Press. 208 pp.
Chapman, V.J. and D.J. Chapman. 1980. *Seaweeds and Their Uses*. (3rd ed.). New York: Chapman and Hall. 334 pp.
Doty, M.S. 1979. The present and future for algal materials. In: Santilices, B. (ed.) *Actas del Primer Symposium sobre Algas Marinas Chilenas*. Subscretario de Pesca, Minist. Econ. Fomento y Recontr. Chile. Pp. 35–49.
______. 1979. Realizing a nation's seaweed potential. In: Santilices, B. (ed.) *Actas del Primer Symposium sobre Algas Marinas Chilenas*. Subscretario de Pesca, Minist. Econ. Fomento y Recontr., Chile. pp. 133–154.
Feinberg, D.A., and S.M. Hock. 1985. Technical and Economic Evaluation of Macroalgal Cultivation for Fuel Production. Solar Energy Research Institute, Golden, Colorado. Rep. SERI/TR-231-2685.
Goff, L.J. and J.C. Glascow. 1980. Pathogens of marine plants. Center for Coastal Marine Studies, University of California at Santa Cruz. Special Publ. No. 7. 236 pp.
Gunther, E. *Ethnobotany of Western Washington* (Rev. Ed.). 1973. Seattle: University of Washington Press.
Hansen, J.E., J.E. Packard, and W.T. Doyle. 1981. Mariculture of red seaweeds. Rep. No. T-CSGP-002. California Sea Grant College Publication, La Jolla. 42 pp.
Hoppe, H.A., T. Levring, and Y. Tanaka. 1979. *Marine Algae in Pharmaceutical Science*. Berlin: de Gruyter. 807 pp.
Korringa, P. 1976. *Farming Marine Organisms Low on the Food Chain*. Amsterdam: Elsevier. 264 pp.
Kramer, Chin, and Mayo, Inc. 1982. Market analysis and preliminary economic analysis for products of the red seaweed *Porphyra*. Report to the Washington Department of Natural Resources. 230 pp.
Miura, A. 1975. *Porphyra* cultivation in Japan. In: Tokida, J. and H. Hirose (eds.), *Advances in Phycology in Japan*. The Hague: Junk. pp. 273–304.
Mumford, T.F. Jr. 1980. Sea culture in Washington and British Columbia: Potential and Practice. In: *International Symposium on Coastal Pacific Marine Life, Proceedings*. October 15–16, 1979. Bellingham, Washington.
Mumford, T.F. Jr., D.J. Melvin, and K. Rossberg. 1983. Pilot-scale mariculture of seaweeds in Washington. In: *Seaweed Raft and Farm Design in the United States and China*. New York Sea Grant Publication.
Neish, I.C., 1979. Principles and perspectives of the cultivation of seaweeds in closed systems. In: Santilices, B. (ed.) *Actas del Primer Symposium sobre Algas Marinas Chilenas*. Subscretario de Pesca, Minist. Econ. Fomento y Recontr., Chile. pp. 59–74.
Noda, H., and S. Iwada. 1978. *Nori Seihin Kojo no Tebiki*. (A Guide to the Improvement of Nori Products). Tokyo: National Federation of Nori and Shellfisheries Cooperative Association. (Translation available from Washington Department of Natural Resources.)

North, W.J. 1979. Evaluacion, manejo, y cultivo de Praderas de *Macrocystis*. In: Santilices, B. (ed.) *Actas del Primer Symposium sobre algas Marinas Chilenas*. Subscretario de Pesca, Minist. Econ. Fomento y Recontr. Chile. pp. 75–128.

Ryther, J.H., T.A. DeBusk, and M. Blakeslee. 1984. Cultivation and Conversion of Marine Macroalgae. Solar Research Institute, Golden, Colorado. Report SERI/STR-231-2360.

Stewart, H. 1977. *Indian Fishing: Early Methods on the Northwest Coast*. Seattle: University of Washington Press. 181 pp.

Tseng, C.K. 1981. Commercial cultivation. In: Lobban, C.S., and M.J. Wynne (eds.) *The Biology of Seaweeds*. Botanical Monographs, Vol. 17. Berkeley: University of California Press. 786 p.

Waaland, J.R. 1981. Commercial utilization. In: Lobban, C.S. and M.J. Wynne (eds.) *The Biology of Seaweeds*. Botanical Monographs, Vol. 17. Berkeley: University of California Press. 786 p.

EPILOGUE

Conte, F.S. and A.T. Manua. 1980. Aquaculture and coastal zone planning. University of California, Sea Grant Marine Advisory Program.

Glude, J.B. 1982. World developments in shellfish aquaculture. Prepared for the Shellfish International Conference, Jersey, Channel Islands, U.K. Glude Aquaculture Consultants, Inc. Seattle, WA.

Hurlburt, Eric F. 1983. Aquaculture in Puget Sound, its potential and possible environmental impact. Washington Department of Fisheries. Draft report.

Island County Planning Department. 1981. Island County regional aquaculture study. Fisheries Production and Systems Planning and Dick-Tracy Associates, Inc. 136 p. + appendices.

Joint Subcommittee on Aquaculture. 1983. National Aquaculture Development Plan. Volumes I and II, Washington, D.C.

Meigs, F.E. 1984. Lending to the aquaculture industry. *J. Commercial Bank Lending*. Pp. 20–28.

National Academy of Sciences. 1978. Aquaculture in the United States: Constraints and Opportunities. A Report of the Committee on Aquaculture, Board on Agriculture and Renewable Resources, Committee on Natural Resources, National Research Council.

Seger, J.L. 1984. State aquaculture plans: the planning process and implementation. M.M.A. Thesis, University of Washington, Seattle. 314 pp.

United States Board on Agriculture and Renewable Resources. 1979. The role of the U. S. Department of Agriculture in Aquaculture. National Academy of Sciences, Washington, D. C. 14 pp.

# Index

Italic page numbers denote illustrations.